BUILD *and* SUSTAIN *a Career in* ENGINEERING

Second Edition

ANINDYA CHATTERJEE

WITH ARTWORK BY GARGI GHOSH

ISBN
Domestic: 978-1-63781-623-3
International: 979-8-89588-399-0

Artwork by Gargi Ghosh

This book is dedicated to
My father, who valued honesty, honor, and originality; and
My mother, who sheltered me from premature responsibility
So that I could follow my dreams.

Contents

Preface to the Second Edition 7

Foreword to the First Edition 13

Acknowledgements 17

1. The Aim of This Book 19

2. The Wage Rate Pyramid 29

3. Actionable Advice 41

4. External Opinion 1 55

5. Learning English 63

6. Campus Interviews 1 77

7. Annual Coin Tossing 87

8. Long Term Value 101

9. Mathematics. 117

10. Campus Interviews 2 123

11. Higher Studies 133

12. Personality and Character 145

13. External Opinion 2 155

14. Failure and Persistence 161

15. Making Your Own Luck 175

16. Looking Forward 189

17. Later Challenges 211

18. The Pyramid Holds the Key 223

19. Some Years Later: A Network of Insights . . . 233

20. Some Years Later: A Philosophy of
    Engineering Design 249

21. The Fundamental Dilemma of Personal
    Ethics 269

22. Self, Workplace, World. 285

Preface to the Second Edition

Many young Indians study engineering, thinking they will get good jobs. But India has too many young engineers and not enough good jobs. A huge wave of young people will soon be looking for work. They will work for less money, and take away jobs from older people. As they themselves get older, new young people will take away their jobs. In a world where knowledge is free and everybody has an internet connection, the advantage of studying in good colleges must shrink. If an employable skill can be learned on the internet, someone will produce an AI tool to teach ten lakh (one million) people that skill essentially for free. When ten lakh job-hungry people compete for one lakh jobs, salaries for everybody will come down. A wave is coming.

I think about these numbers, so I wrote this book.

What is to be done?

I believe that much can be achieved with a plan, with effort, and with time.

If you reach age 40 with the same skills that 22-year-olds of that time have, then you are likely to be laid off because those 22-year-olds will work harder for less money. There is only one way to stay out of such rising water: keep climbing. If you do not wish to climb, I cannot help you. If you wish to climb, you need a long-term plan. I offer a plan.

You may already have a plan. For example, your plan may be to study artificial intelligence and machine learning (AI and ML). These are certainly growth areas. But there is no escape from statistics. Even if a field has one million jobs, if there are two million young graduates in that field then one million must remain unemployed. A time may come when too many Indians have taken good courses in AI and ML. Would you like to be 40 years old when that time comes?

In any case, the older or core engineering subjects like electrical, civil and mechanical engineering continue to be taught widely in India. Luckily, India's core engineering problems are far from solved: not just in companies, but also in many aspects of daily life, in homes and communities. Where needs exist, new opportunities may emerge. Moreover, the core engineering subjects, perhaps accidentally, teach a lot of valuable transferable skills. You may have heard of mechanical or electrical engineers who are economists or computer scientists or novelists or entrepreneurs.

Regardless of which field you go into, this book has something for you.

There are already many people offering excellent advice on shortcuts, or quick ways, to good careers. The problem is that many people adopt that advice, because shortcuts have low entry barriers. Soon the advantage disappears. This is why I prefer a long-term plan. The "long term" part is the entry barrier. Most people are not patient.

Much of the book is written for young engineers. Near the end I discuss higher level abstract skills that you may appreciate only over a period of ten years or more. If you are a beginning student, think of them as a glimpse of things that lie far ahead on your road. Do not worry if older people know those things and you do not. This gap between older and younger people is why *you* cannot replace those older people today, but it is also why *you* will be difficult to replace several years later. These chapters draw from my own experiences, and may be most useful to students of applied mechanics (mechanical, aerospace, civil, and automotive). However, the principles apply to other fields.

Contributions from two friends are included in individual chapters. These people have had careers that differ greatly from mine, and their inputs help to complete the book.

I have had many discussions with colleagues and friends over many years. They have influenced my thinking. However, except when stated otherwise, the book presents my beliefs and opinions. Errors and omissions are mine as well.

A few chapters in this book have drawn from some answers I wrote earlier on the internet platform *Quora*. Most of the chapters have been written directly for this book.

Finally, this is a book of opinions and advice. You may not agree with me on all points. But if even a few of the chapters resonate with you, give you confidence, or prompt you to become stronger and to reach higher, then my book will have been successful. Through my chapters or otherwise, if you build and sustain a satisfying career in engineering, then my goal will have been served.

Important changes in this edition of the book, compared to the first one, are in the exercises included at the ends of chapters and in the new chapter on ethics.

IIT Kanpur
January 2024

Foreword to the First Edition

Engineering education is concerned with imparting knowledge in engineering and technology practices. Engineering innovations form key elements for the sustainable growth of any country. Technical human resources with appropriate skill sets contribute immensely to the creation of wealth. Given its importance for economic growth, engineering has been, historically, attractive to young students.

With more and more students aspiring for engineering education, beginning in the late nineties, there was a proliferation of engineering colleges all over India. Today, India has more than 4000 engineering institutions and produces approximately 1.5 million engineering graduates each year. India has emerged as a major player in engineering education in terms of numbers. The expansion of engineering colleges, however, has not been accompanied by the quality of education. Despite the large number of graduates being produced every year, according to one estimate, 50% of Indian engineering graduates today are either unemployed or under-employed. The quality of engineering education in most institutions requires improvement and the skill sets acquired by the graduating students are considered to be inadequate by the industry. Several core engineering industries are reluctant to hire and often blame it on the low quality and skill sets.

Apart from unemployment, many graduates are under-employed in the sense that they are working in companies and organizations where the technical skills required are not commensurate with their engineering education. Specially in the beginning of the twenty first century, India witnessed a boom of BPO and services industry which employed large number of engineering graduates with job profiles which did not require real engineering skills.

While there are a large number of core engineering jobs in the country, some sectors also suffer from lack of focus on innovations, IPR creation, product design and manufacturing. This has been especially true in electronics & telecom, healthcare & medtech etc. Recent economic disruptions brought by the Covid pandemic situation have re-emphasized the role of engineering innovations. For economic prosperity, we need a sustainable and thriving eco-system of R&D driven industry which can harness the potential of our young graduates. At the same time, we need well-trained confident engineers who can leverage the emerging market scenario and possess the skills required by the industry. This reality of Indian industry (at least in some sectors) coupled with the large number of engineering graduates with inadequate training has led to a vicious circle.

This book by my colleague Prof. Anindya Chatterjee attempts to address one part of this problem, which is how engineering graduates can "build and sustain a career in engineering." Our engineering graduates need to not only hone their engineering knowledge, they also need to understand the practical realities of life and improve their soft skills. Above all, engineering graduates need to develop confidence in addressing real-world challenges. These have been articulated in the book in an extremely lucid style. The book has clinically analyzed the problems faced by engineering graduates aspiring to get employment in industry and has offered some very actionable advice.

While each career requires an employee to have appropriate skill and confidence in handling real-life challenges, these skills need to be integrated with an engineering temper for a technical job. I hope this book will inspire and motivate our young graduates to not only become employable but also become confident engineers who can then lead the industry in their careers.

Once you get employed, you need to sustain and furthermore drive the industry eco-system to create a world of opportunities for others. This creates a positive feedback leading to a sustained economic growth. I hope this book will prompt you to develop your strengths in that direction.

Abhay Karandikar
Former Director, IIT Kanpur

Acknowledgements

Several people have read earlier writeups I have used here, or read the book, and made useful comments. They are: Sarmila Banerjee, Dipankar Basumallick, Neeraj Biyani, Devlina Chatterjee, Riju Chatterjee, Joseph Cusumano, Chiradeep Datta, Pramod Kumar, Mohit Law, A. G. Menon, Atanu Mohanty, Anjan Ray, Suresh Reddy, Sanjoy Sanyal, Ishan Sharma, Shashwat Srivastava (who also drew a picture), Sankalp Tiwari, C. P. Vyasarayani, Asok Mallik, Sarma Rani, and G. Ananthasuresh. I thank them all for their time and help. I especially thank my wife, Devlina, who nudged me into writing this book instead of just talking about these things.

CHAPTER 1

The Aim of This Book

First, the serious news. Let us get it over with.

India has 1.42 billion people and our median age is 28.7 years (numbers from 2022). That means 710 million people under age 28.7. Unemployment is high. For the next generation of Indians, the competition for good career outcomes will be harsh. If you are a young Indian, this is a good time to recognize the scale of your coming battles.

710 divided by 28.7 equals about 25. Roughly speaking, we have about 25 million 17-year olds, 25 million 18-year olds, and so on. We have about as many 18-year olds as Australia has people of all ages combined. Meanwhile, all of Infosys and TCS together (for example) have about one million employees.

The wall is 120 kilometers long.

In this book, at the cost of a minor inconsistency in the naming of numbers, I will sometimes use "million" which means 1,000,000 and sometimes use "lakh" which means 100,000. I could use 0.1 million instead of 1 lakh, but I think in some cases using "lakh" helps in connecting with the number. Also, 10 million or 10,000,000 is 1 crore.

Sometimes we read numbers without really connecting with them. Let us construct a mental image to help us. Several years back, the Indian railways had announced 1.2 lakh jobs, and 2.4 crore people had applied (i.e., 24 million). That means an average of 200 applicants per job. That factor of 200X increased in subsequent hiring efforts, but 200 is a good number for this example.

Think of a long wall with many doors along its length. There is one door per meter along the wall. The doors are shut, but there are people waiting outside. Outside each door, on average, there are 200 people waiting in line. One meter along the wall, another door, another 200 people. At some point, each door will open, and one out of these 200 people will be allowed to enter. How long is the wall? It is 120 kilometers long. At Indian driving speeds, it may take you two hours to drive past these people. If you are one of the job applicants and discover that you have been standing in the wrong queue, it may take you an hour-long taxi ride to reach the correct queue. This is one glimpse of the unemployment problem in India.

You may agree that a long-term plan is needed. That is what this book is about. You may think, too, that the solution lies in better education. Maybe it does, but the matter is not simple.

India's education needs are poorly served. Most people in India cannot find seats in good colleges[1]. As a result, many of them study engineering, thinking that it will get them jobs. Every year about a million young Indians compete for a few tens of thousands of "good" engineering seats. Many others study in less reputed engineering colleges.

India has thousands of such less reputed engineering colleges. For typical students, these colleges offer poor career outcomes. The problem is two-sided. On the one hand, many engineering graduates are said to be unemployable. On the other hand, even if these graduates *were* employable, it is not clear who would employ them. There are jobs in the IT services sector, it is true, but there are not nearly as many jobs as there are graduates. There are also many jobs in various

---

[1] I grew up in Jamshedpur and Dhanbad. The best degree colleges of such towns are less sought after than the good colleges of New Delhi or Kolkata, for example.

small engineering or manufacturing companies, but the technical demands of the jobs are low and salaries are small.

Every year, several lakhs of engineering graduates appear for the Graduate Aptitude Test in Engineering or GATE, for postgraduate education as well as a small number of jobs with public sector units (PSUs). Most people who appear for the GATE spend a year or more in preparation, but end up with neither good admissions nor PSU jobs. Many others prepare for a big exam conducted by the Union Public Service Commission or UPSC. UPSC success rates are tiny. I will write later in the book about a candidate who tried this exam and failed.

Clearly, we have a big problem. Solutions for this big problem lie at many levels. For the nation, these are deep and big issues of policy. For engineering colleges, the challenge is one of trying to train people to be more useful twenty to thirty years from now. For students as a group, the search is for areas where they can contribute and earn decent livelihoods. For the individual, it is a matter of developing superior and lasting professional value. This book is written for individuals.

Let me set apart the question that drives this book.

How can you improve your probability of achieving better long-term career outcomes?

There are some well-known shortcuts. Make better eye contact; practice a firm handshake; read about current topics in business magazines; do some research on the company before you go for an interview; learn coding; take a course on artificial intelligence; and other things like that. These are valuable things, and there is a large amount of good advice on it. A Google search for

"how to ace an interview"
(with quotation marks) fetches a huge number of results. This book is not about such shortcuts.

Of course, I like shortcuts. They are efficient. The problem is that shortcuts have low entry barriers. Even as *you* take them, a million others take them too. They take them just as easily as you do. Not taking them puts you behind, but taking them does not put you very far ahead.

You must do more. If you are willing to put in time and effort, and to take responsibility for your own competence, then you are ahead of the game. You will develop a lasting value proposition that may serve you well for a long time.

The required persistence is the entry barrier. Your willingness to do the extra work is your protection. Many of your contemporaries will turn out to be lazy or shortsighted. If they let ten years slide by while *you* move forward with a plan, most of them will not be able to catch up with *you*. Statistically speaking, they will have poorer outcomes in the long run[2].

Let me put it another way. Who is in the top few percent of the engineering population? It depends on the criterion you use. In the short run, people who do well on the JEE Mains may be considered to be in the top 2 percent. But it is not clear who, in the long run, will have satisfying and stable careers; who will change lives for the better; who will find fulfillment and not just a good bank balance. There is more than one way to choose a top few percent. Many people are impatient and do not look far ahead.

This book is a call to the top few percent of engineers in terms of patience, the ones who want to look far ahead, the ones who are interested in building genuine multi-dimensional long-term worth. I believe there will be a place for such people in this world.

---

[2] Loosely speaking, let me suggest: success equals *talent* plus *labor* plus *strategy* plus *luck* plus *social advantages* minus *competition*. Talent is something you are individually born with. Social advantages (e.g., from family background or contacts) are unequally distributed within the population. Competition is high for all. That leaves labor, strategy, and luck. This book is about these last three.

Exercise 1.

Let us end this chapter with a small exercise. I wrote above about a railways job search success rate of 1 in 200. It is tempting to think of those applicants as "them" and not "us." We think we are not like "them." Let us try to connect better with the mercilessness of probability. You will need a pencil, paper, and 30 seconds.

Write down three numbers (integers) between 0 and 200, not necessarily in increasing or decreasing order. Choose afresh, and randomly, each time. Write them down as (a) first number, (b) second number, (c) third number. Now compare with my randomly chosen numbers below[3]. The probability of having guessed all three of my numbers in the correct sequence is one in 8 million. If you did it, you just won a lottery. If you think you can win repeatedly in this way, this book is not for you.

The probability that you guessed *all three* incorrectly is $0.995^3 = 0.985$. If 1000 people play the game, we expect about 15 to guess at least one number. If you guessed one, congratulations. Your success rate is comparable to the national success rate of getting into IITs.

Did you guess two numbers correctly? The probability of guessing two is less than one in ten thousand. Think of a long, overcrowded train. 20 compartments, 200 people per compartment. Lots of people in the train

---

[3] (a) 163, (b) 181, (c) 25.

are standing. We are talking about something like *one* winner from *three* such trains.

The risk of a failed career is a big one. Where there is risk, we must think in terms of probability. The aim of this book is to help you think about career planning from a probabilistic viewpoint. What plan is likely to work? Are your chances one in ten? One in three? Is there a plan that can improve your chances from one in ten to one in three?

Read on.

CHAPTER 2

The Wage Rate Pyramid

We are engineers. We think in terms of models. Let me begin with a simple pyramid model for the job market. This model will help you think about the forces that will shape your career. This is the most serious chapter in this book. It may help you see why I think it is essential to develop a strong and long-term value proposition if you want to build and sustain a career in engineering.

First, consider a time of low unemployment. If there are more jobs than applicants, then some employer somewhere is left who still needs an employee to make profits. That employer must either forgo profits or offer higher wages to attract an employee away from another employer. This can lead to a spiraling rise in wages until they take up a sizeable share of profits.

The opposite holds in times of high unemployment. If there are more applicants than jobs, then someone remains unemployed. The unemployed person agrees to work for less money, which can lead to some other person being laid off. This process may continue until all wages for that specific service (whether grass cutting or machine design or IT services) are driven to a lower equilibrium. In this way, even moderate unemployment within a particular job, or within a well-defined skill set, can lower wage rates by a lot.

Let us be clear on this. I am not merely saying that free market wages go up or down depending on whether *overall* unemployment is low or high. What I am saying is more serious because it applies to each specific skill set. Free market wages go up or down for each specific skill set based on the relative excess of workers who offer that skill set. *They go up and down by a lot.*

The narrower the skill, the greater this effect. To help fix the idea in your mind, let me offer examples from food. Indians eat *daal*, of which there are several varieties. Some level of substitution is possible. So if *arhar* is in short supply and its prices go up, people may buy *moong*. Due to this, *daal* prices may go up and down together, even if not identically. But there are food items where substitution is less easy: onions, tomatoes, and lemons, for example. You may recall times when one of these items became very expensive or very cheap, with a much smaller effect on the prices of other items.

Returning to wage rates, if you are a young Indian, then I hope you will understand that the consequences are serious. Let me offer an example with hypothetical numbers. Imagine that you are an employer. You own a machine that produces some component. If you hire an operator, your hourly profit will be 200 rupees minus the operator's wages. There are four others like you, making five employers in all. But only four operators apply for work. The other four owners hire them, each at 100 rupees per hour. They are now making profits of 100 rupees per hour, while you are making zero. You offer one of those operators 120 rupees per hour to work for you instead. Soon, the operators are all earning 120 rupees per hour, and one employer is still missing out on a possible profit of 80 rupees an hour. That employer then offers wages of 140 rupees. In this way, wages rise towards 200 rupees. Now imagine the opposite. There are four employers and five operators looking for work. Each operator needs at least 40 rupees hourly to support his or her family. Four operators are hired for 100 rupees per hour. The fifth operator needs money and offers to work for 80 rupees per hour, getting one operator laid off. Soon, all operators are earning 80 rupees each, and one operator *still* remains out of work. Wages then go down further, towards 40 rupees[4]. It can get worse than that when there are many jobs and many more applicants. For example, if it is possible to overwork employees in a time of unemployment, then some employed people may get laid off and the remaining ones may be asked to work harder. In terms of the above example, it may happen that instead of retaining 4 workers for 4 jobs, one further worker is laid off and the remaining 3 workers are asked to do 1.33 jobs each[5].

Perhaps you think this is *not* how the private sector job market works. Maybe you think that wages represent the true value of work done, whatever *value* means. If so, think about how domestic cooks are hired in India. Their wages are not based on what value eating has,

---

[4] The free market improves efficiencies. Suppose there are 5 employers and 4 workers, and one inefficient employer can earn only 140 rupees per hour minus operator's wages. Then wages move to 140 rupees, and that employer must either become more efficient or leave the business. Conversely, when there are 4 employers and 5 workers, suppose wages move to 50 rupees. Now a worker who cannot support his family below 60 rupees is the inefficient one. He must either live on less or work longer hours. It is arithmetic, not ethics.

[5] At the time of writing this, two recent news items came to mind. One was Elon Musk buying Twitter (now called "X") and laying off about half the employees and asking the remaining ones to keep working. The other was Narayana Murthy of Infosys saying that young people wishing to build careers should work 70 hours a week.

which is high. Nor are their wages based on the value of the employer's time or health. The cook's wages are basically determined by the wages of other cooks in the neighborhood. For another example, think of schoolteachers. Schoolteachers hold our society's future in their hands, and the true value of their work is high. Then why are they almost uniformly underpaid and overworked? Because applicants outnumber vacancies in their part of the job market.

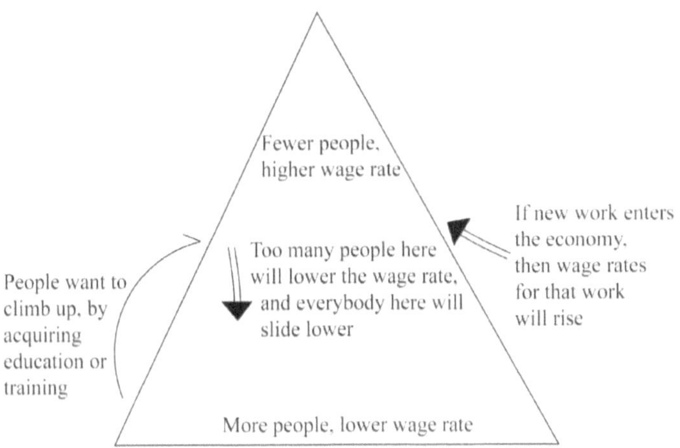

The wage rate pyramid.

The hierarchy of wage rates in the job market resembles a pyramid. It is a dynamically changing pyramid. A few people at the top earn high wage rates. Increasingly larger numbers earn steadily lower wage rates as we move down the pyramid. Since the pyramid is defined using wage rates, each level of the pyramid may include many services or skill sets.

People familiar with basic IT services salaries in India have already seen the dynamics of the wage rate pyramid. Initially, when many jobs came to India from the west and there were relatively few people in India fit to be hired for those jobs, wage rates rose to high levels. Soon, seeing those high wage rates, lots of young people acquired IT skills. The wage trend was reversed. For a decade now, the entry level wage rate for basic IT jobs has not moved much in nominal rupees and thus been eroded by inflation[6].

Based on the above I suggest that, for any well-defined skill set, the middle ground is narrow and perhaps unstable. Too few people available to do the work tend to drive wages far upwards; and too many people available to do the work drive wages down to subsistence levels. The wave of young people entering the workforce today can expect trouble if they offer well-defined skill sets that are also offered by even younger people entering later. Young workers today

---

[6] If wages in rupees were to increase at a rate of 6% a year to match inflation, then in 12 years they should increase by a factor of 1.06^{12} or about 2. Entry level wages of about Rs. 3 lakhs per annum from 12 years ago should now be Rs. 6 lakhs per annum. But they remain at around Rs. 3 lakhs per annum. Wages, therefore, are now worth half of what they used to be. Why? Acceptable applicants outnumber vacancies.

need to steadily build their value proposition: to rise above the young people of the future. Think of the young people behind you as a rising tide. You have two choices: climb, or get wet.

When a specific new service or skill set is required in large volumes (like the initial stages of the IT boom in India), new work enters the pyramid and wages for that service go up. Each well-defined service that attracts a high wage rate soon attracts new workers from below who learn the skills required for that service. Eventually, when there are too many people both willing and able to provide that specific service, the wage rate for the *entire* service is pushed down. All providers of that service, both old and new ones, slide down to a lower level in the wage rate pyramid, as seen recently in the Indian IT industry.

For a wealthy western economy, sending work overseas may directly reduce their local wage rate for some tasks, while increasing the wage rate elsewhere. The Indian IT industry has benefited from this for several years. However, AI (artificial intelligence) and ML (machine learning) may now take some of that work away and give it to computers, and then many Indian IT professionals may slide down further[7].

The pyramid also welcomes and includes new services and skill sets within itself. If people keep offering profitable new services, new and high wage rates will appear. A job that requires a skill set which is not easy to mass produce may be safer from collapsing wage rates.

Let us consider some examples. Times are not very good now for a fresh mechanical engineer from a small college seeking a core job in India. Mechanical engineers have traditionally managed factories and manufacturing processes, designed components, worked in power plants, maintained factory equipment, and so on. Today there are too many graduates every year, from many Indian colleges, who can be trained quickly to perform these tasks at the level needed by most Indian companies. Wages for such work have therefore fallen and may not rise much soon.

Occasionally, a mechanical engineering team may face a problem they cannot solve. They may then engage a consultant who helps them find a solution.

---

[7] Roughly speaking, AI/ML turn creative jobs into clerical jobs, and remove clerical jobs altogether. Indians in large numbers are rushing into these subjects. If and when there is a glut, wage rates must fall.

Such consultants are difficult to mass produce, and may attract a high wage rate for some more years.

Some companies develop new and different designs on a regular basis and may prefer to hire from higher ranked institutes[8]. As the employer gains experience, and as more colleges produce capable graduates, students of such higher ranked institutes will have to either raise their own value offerings or accept lower wages.

Some careers have a steady learning curve. Every year on the job makes one harder to replace. Examples of such people are surgeons, nurses, plant maintenance engineers, and managers of large field projects. In such jobs, one's competitors are contemporaries. Younger people tend to remain behind on both skills and wages. Such jobs may survive the next generation's wave of job seekers.

In contrast, in many other jobs, passing years do not build greatly superior skills. Younger people may soon match older ones, and machines (or AI) may eventually take over from people.

Relatively few jobs guarantee both a decent wage rate and continued employment. Examples are permanent

---

[8] When a task is open ended, precise evaluation is difficult. Then who is to judge whether the employee's work is good or bad? In such cases, the institute's reputation is used as a proxy for the employee's quality. Look up *job-market signaling* for more on this idea.

jobs in Public Sector Units or PSUs in India. Here, the difficult part is getting the job. Subsequently, PSU employees are protected from the dynamics of the wage rate pyramid. Recent years have seen high competition for PSU jobs among young Indian graduates. Since the total number of such jobs is small, getting such jobs is not assured for even the top few percent of the young engineers that India produces every year.

A final example, and a completely hypothetical one. Imagine that a wonderful computer-brain interface is invented so that *everyone* in the country is given exactly the correct amount of knowledge needed to get an MBBS degree, completely free. We instantly have a country where *everybody* is a doctor. What will happen to doctors' fees for routine advice? I think they will fall to zero. That fall to zero is the downward slide of the basic medical skill set on the wage rate pyramid. Of course, you may sometimes need more than routine advice. If a loved one is seriously ill and you decide to consult a doctor, how will you choose one? I think you will prefer a doctor who has a track record of treating many complicated cases, and you will pay whatever fees they charge. That doctor will have risen in the wage rate pyramid.

In this hypothetical example, when everybody has basic medical knowledge, the fees paid for basic medical advice will become zero. Yet, even in that situation, doctors who have long and broad experience of treating complicated cases will still be able to demand high fees. Agreed?

Replace everybody with *too many*, and replace medicine with *engineering*, and the situation is no longer hypothetical. That is why I have written this book.

Exercise 2.

Let us end the chapter with an exercise.

First consider a specific skill like using a mouse-driven software package to make engineering drawings, or finding the roots of a cubic polynomial using a calculator, or using some existing ML software for an application where the training data is available in an Excel file. I suggest that these skills can be learned online.

Now imagine that you are a rich person, and you just want to help India's millions of poor and unemployed people. Imagine that you hire a few top-notch computer engineers, and they develop an AI platform which adjusts to individuals' learning needs and speeds, gives them graded assignments, and slowly but surely trains millions of people in such a skill.

Next, imagine some skills of a different kind, like riding a bicycle, driving a car, welding steel, doing surgery, doing higher level mathematics, conducting failure analysis of engineering systems, or managing teams of workers on construction sites. I suggest that these skills are harder to learn online. Think about how AI can be used to train people to do such work. You may agree that AI can help, but the main burden of learning still lies heavily on the learner. You actually have to do the work in a place where the computer cannot reach easily.

By now you may agree that not all skills are equally amenable to being taught online to millions using AI.

Now think of two teams, R and S (for "risky" and "safe"). Take 5 minutes. For every job-related skill you can think of that is at risk of being taught easily by AI, give the R team one point. For every job-related skill you can think of which is somewhat safe from AI-based teaching, give the S team one point. Which team won?

Will this exercise affect your career planning?

CHAPTER 3

Actionable Advice

The world is filled with people competing to offer you advice. I am one of them. You must decide whether the advice is any good. This chapter is on evaluating advice.

To be useful, first of all, advice needs to be *actionable*. I do not mean actionable in the legal sense, but in the sense of *possible to act on*.

Let me give an example of what I mean by actionable. Once a senior person from the management of a leading automobile research organization visited our campus and gave a talk on their activities. He had many slides, and spoke at length about the history, size, and scale of his organization. When it came to the work they did, he had many bullet points and spoke for several minutes. More than thirty minutes into the talk, however, with nothing actually technical so far, he was speaking of improving the mileage of vehicles in a general sort of way.

I finally requested him to be specific. No vehicle has a mileage improvement button that you just go and press. Mileage improvement is a result, not an action. The tasks involved may include first running the engine under a variety of torque and speed conditions, with a variety of fuel injection strategies based on certain numerical parameters (perhaps timings, volumes, pressures, …), and measuring the engine efficiency under each load condition for each set of parameter settings. How many such settings are studied? How long does such a project typically take? How are the optimal conditions identified? How are they implemented in the vehicle (with passive mechanical devices, or with electronics)?

I made the social error of pointing out to him that the task, "improve the mileage of this vehicle," was too vague. We were in a technical institution, and we would be interested in hearing actual details with a clear list of actions. Sadly, all I did was offend the speaker. He told me that these were not easy matters, that he had given many talks to many people, that he had 250 slides with him right then, and that he did not have time to show them all to us. We did not get anything actionable from him in that talk. All we knew at the end of the talk was, "They are big, and they do things such as improving mileage." But we knew these things before the talk.

I agree that improving mileage may be a perfectly actionable task for someone who is an expert in that field, e.g., one of the experienced engineers of that organization. If I worked for a car maker and I went to his organization and asked for their help in mileage improvement for my company's products, the conversation would be meaningful. But, for the audience of that talk on that evening, "improving mileage" was not actionable.

Now for a hypothetical example. Consider yourself, a young student, presumably studying hard. Your favorite uncle comes along and says, "Improve your health."

Your uncle's instruction is not actionable unless you already have a prior conversation going on about this issue. Consider this instead: "You are working very hard, and I am afraid it may affect your health. Add a little extra protein to your diet, make sure you sleep at least seven hours each night, drink a bit of extra water each day, and try to go for a run or at least a brisk walk every evening." That is actionable. You may or may not follow his advice, but there is little doubt about *how* to follow it in case you want to.

Several years ago, I was watching a senior colleague advise a group of students on how to give a technical seminar. "Your slides must be pleasing," he told them. "You must be confident. You must be prepared for all kinds of questions." I find such advice to be *not* actionable.

More actionable advice is possible. I give the following mundane, but actionable, advice to my students about technical seminars.

1. Select the background color and the text color in advance for good contrast and low eye strain. Use the same combination for all the slides, except maybe with one or two exceptions for emphasis.

2. The largest and smallest fonts in any slide must not differ greatly in size. The same holds for your entire set of slides.

3. If it takes you more than one minute to read the text on a slide at a normal pace, then your slide has too much text. Cut your text, or use two slides.

4. If you don't want me to see something, don't put it on your slide. If you must put it on your slide but you don't want me to get distracted by it, say so clearly the moment you show the slide, before I get distracted.

5. Avoid large tables of numbers. If you must put in a large table but only want to discuss a few numbers,

make those numbers **bold** and tell the audience your intentions. For example: "Sorry about the large table of numbers. I will be focusing on the numbers written in boldface."

6. If you put in graphs, make sure your axis labels are visible and your lines are thick enough to be seen clearly.

7. If you show a complicated picture that is important to your talk, offer a visual cue. Pause visibly, or smile for a second, or nod at the slide: something like that. Then say something like, "Let us go over this figure slowly." This helps your audience stay relaxed.

8. Your laser pointer's dot of light should be steady on the screen. Nobody likes a randomly jumpy point, and it makes you seem nervous as well. Look at the dot as you talk. Turn slightly, as a cue to your audience to look at it too.

9. If you do not want your audience to look at the laser light dot, turn it *off*. Definitely don't keep it on and wave it around all over the room while you look elsewhere and talk about something else.

10. A rough time budget is one minute per slide. But practice in advance and see how much time you need. At any instant, if you glance at the clock and look at your current slide, you should know whether you are keeping time.

11. Identify in advance one to three slides in the second half of the talk that can be sacrificed. Then if you are running short of time, just say, "These slides discuss such and such, but I will skip the details now and return to them if there are questions." This can help you recover precious minutes.

12. Eye contact is important. Let your eyes move to various members of the audience. Do not just look at a small, fixed set of people.

13. The loudness of your voice is important. This issue requires practice. In a moderately sized room, the person in the first row should not feel that you are shouting, and yet the person in the last row should be able to hear you. Ask if they can hear you at the back.

14. Make a video of yourself giving a practice talk. Watch it more than once. Look for places where your speech falters, then plan the words you will use there to have them ready in your mind. Look for repetitive or distracting movements that you might be making unconsciously (speed up the video to see them more clearly).

The above suggestions are, in my opinion, actionable. You may or may not use them, but you know what to do if you want to use them.

Why do some people offer advice that is not actionable, and why do other people listen to such advice with care? My theory is that in our minds we have processes of both logic and emotion. When there is something that we cannot resolve with logic, we feel some stress; and then we may welcome the relief that an alternative emotional process provides.

Let me offer an example.

Once, as a young boy, I was attending a wedding ceremony. It was crowded, noisy, hot and humid; there were extra lights and pedestal fans; there was smoke from the ritual fire; and people were moving about. I stopped for a minute to chat with my uncle. My uncle's son, then a toddler, came up and said he was thirsty. There was a table about fifteen feet away with several glasses of water on it. My uncle told his son, "Can you go to that table and get a glass of water all by yourself? You will have to be very careful. There is a wire over there on the floor. Can you be careful?"

Be careful.

The child's face changed. He became solemn, looked at the wire, and promised to be careful. Then he set out, put each foot down very deliberately, stepped over the wire with exaggerated caution, turned slowly to look back at his father (who nodded), kept walking with great care, got himself his water, drank it, put down his glass, and walked all the way back with equal care. He was pleased with himself. I realized only later that the child did not really have any idea *how* to be careful except to not trip over the wire. But the emotional satisfaction he got from taking exaggeratedly slow and deliberate steps made him feel good.

Revisit, now, the technical seminar giving advice of my senior colleague: Your slides must be pleasing. You must be confident. You must be prepared for all kinds of questions.

I imagine that the student being offered this advice is already feeling some stress at the thought of giving a technical seminar. He is insecure both personally and academically. Now, he first imagines his slides being pleasing, and that pleases him. He imagines himself being confident, and that pleases him further. He finally imagines himself easily answering all kinds of questions, and feels very good indeed. He likes the final feeling, decides to feel this way after his seminar, and forgets to check if the advice is actionable.

What about the person offering the advice? What are *his* motives behind giving advice of such dubious utility? I can only guess. Perhaps he is just an older version of the student, and never noticed that his advice was not actionable. Perhaps he is just expressing his own emotional response to unsatisfactory talks of past students ("ugly slides, nervous demeanor, unable to answer questions"), and feeling some relief from these utterances. Maybe he has psychological motivations which have more to do with him than with either the seminar or the student.

Advice that is not actionable is common on the internet. I once tried a Google search for advice to startups and visited the very first page that came up. It was an article titled *8 Essential Tips for Small Business and Startups from Expert Entrepreneurs* by Chelsea Segal. The tips were:

1. Take risks and be willing to fail.
2. Never stop networking.
3. Learn your niche.
4. Be a consummate student.
5. Don't worry about your wallet …
6. … but try not to go broke.
7. Be flexible and listen to the market.
8. Take care of yourself.

I am not an expert entrepreneur. Perhaps Segal's article has things of value based on real experience. But I must admit that at least some of these tips leave me unclear on the *action* required. I would, in contrast, know what to do with something along the following lines (an inexpert non-entrepreneur's tips, if you like):

1. Find out how to start, register, and close a company. Find out both legal and financial requirements.
2. Identify your partners, if any. These must be people you trust completely.

3. Propose a product or service that you will provide. Then discuss this in great detail with your partners. Iterate on your proposal until you are convinced that it is your best choice, as below.

4. Is there someone that can stop you from offering this product?

5. Is there enough latent demand in the market to generate the volumes you will require to survive?

6. Is there someone who can move in easily after you establish a market, and take your business away from you?

7. Decide on how work and profits will be shared among partners. (I have heard that the profits must go, in decreasing order, to sales, then operations, and then accounting and other control.)

8. Start work full time. At this stage, your company is your top priority. This is not a job with fixed hours.

9. Look for someone to fully or partially finance your venture. Selling a part of your business early covers your downside, and lowers your risk.

10. As you start getting results, think about reaching out to your market. If you have powerful competitors, you need to wait longer until you announce yourselves. If you have a big initial advantage, start marketing conversations earlier.

In the rest of this book, I will try to offer advice that is both useful and actionable.

Exercise 3.

It is time for an exercise.

Imagine that you lead a small team of engineers. Your team has worked on a project. One team member has prepared a PowerPoint presentation. Your entire team has watched the presentation. You find the main message to be unclear; the statement of the contribution of your group to be unclear; the slides to be overcrowded; the font sizes to be sometimes too big, sometimes too small, sometimes both on the same slide, sometimes in a chaotic variety of colors; the project's goals to be incorrectly stated; and the conclusions to be incoherently listed.

Be bad. Using bullet points, make up six items of feedback you can give to your team members that are seemingly valid, yet *not at all actionable*. Write them down and imagine a future annoying boss giving you that feedback.

Make up some sentences whose action items are not clear. I can offer a few to get you started.

The energy in your slides is not harmonious. Your slides must be more confident. *Yaar, mazaa nahin aaya.* Your slides should be more impactful. Clearly, you do not understand psychology. Your slides gave me a headache. The color scheme is a bit immature, isn't it? Your delivery shows you have no experience. How can you expect to get the contract with a presentation like that? Show me something more effective. Put in the hard yards. Show me that you have what it takes. Walk the talk. Believe in yourself. How can you have so little faith in our product? This is not good enough.

CHAPTER 4

External Opinion 1

Chiradeep Datta and I were batchmates in college. We had neighboring rooms in our hostel, and remain close friends. I became a professor. He joined industry and did real engineering within India. He has worked for GAIL, then Reliance, and then AG&P. When he was with Reliance, he was (among other things) head of field operations in a project that laid a 500 km long pipeline for carrying liquid ethane. Liquid ethane requires rather high pressures: it is not as easy to handle as LPG. Think of such a project in India: issues of geography, culture, acquisition, technical specification, reliability, inspection, commissioning. Think, if you like, of just the number of pipe welds involved: seams, girths, in the plant, in the field, pressure testing section by section, for a 500 km long pipeline.

Datta's view of what constitutes value in an employee is, clearly, shaped by experiences vastly different from mine. Here is a short piece he was kind enough to write for me. The quoted portion below is from him. Subsequent comments are mine.

Any institute will be teaching students who will, in future:
1.   Work for entrepreneurs, business houses and industrialists.
2.   Be entrepreneurs themselves.
3.   Remain in academia in pursuit of research and excellence in teaching.
4.   Branch off into some other field like engineers into management, doctors into administration, sportspersons into journalism, and journalists into music.

For this discussion, I am limiting my comments only to Category 1, where students learn in the institute to go on and create wealth for others.

The existence of this category of students automatically presupposes that there are people (from private and public sectors) who are in the business of creating wealth and those people require certain human resources with defined skill sets in their pursuit of making money.

This brings us to a discussion of how the wealth is created and what kind of human resources will be required in creating that wealth.

For wealth to be created, business owners need to address some kind of a need for which someone would be willing to pay; and should have a process in place to meet that need with a saleable product or service.

For that, the business owner needs the following:

1. A mechanism to identify new and unstated needs; for which he or she needs a business development team to study market needs and understand trends, etc. He or she would then require innovative people who can develop products to meet these needs.

2. If a business owner operates within an envelope where needs are already identified and products are widely known, then the business would face competition from other entrepreneurs. To sustain the business, the owner will have to be cognizant of the little changes in market requirements, the risks that the business faces, major threats from changes in the business environment, and the artificial roadblocks that may be created by business rivals.

3. He or she also needs an efficient operations team focused on lowering costs and improving safety (thereby lowering unforeseen costs due to accidents, product losses and environmental damages).

4. He or she also needs a good finance team who can ensure availability of low cost working capital and take care of business risks like currency and energy price fluctuations, and assessment of market demands so as to optimize inventory carrying costs

and prevent situations of product stock outs or abnormal inventory buildups.

5. The business should also comprehend consumer difficulties. For example, a company selling seeds to farmers prior to sowing season should understand that the farmer may need to borrow money. For ensuring sales, the business may need to tie up with a financing company which provides loans to farmers at comfortable terms and in return the farmers would prefer to buy seeds from this company.

Thus, a business house would prefer employees who have one or more of the following varied qualities or skills.

1. An employee will be preferred who is capable of need identification and has the capability to convert that need into a product or a service. Such an employee should be (a) good innovator and (b) also have the tenacity to develop and test a fully functioning product.

2. An employee will be preferred who is capable of process risk assessment and risk mitigation. Employees who are safety conscious will do well.

3. Employees who understand the concept of operations or production lines and are good at grasping operational risks would also be highly employable.

4. Employees who combine skills in operations or supply-chain with financial risk management will be a hit. Those capable of designing financial tools to manage risks will get high paying jobs.

5. Finally, people who are good at interacting with consumers and intimately understanding their

difficulties, which information can then be used to package their products with add-ons, will be employable.

For engineering students, those good at reliability studies, and those who have a flair for understanding efficiency of systems, will always endear themselves to employers as they can innovate and drive down production costs.

However, those who can identify implied needs and bring out new products will remain ahead of the rest, as employers will be using them to beat the competition.

Financial risk managers who understand engineering, and marketing managers who understand consumers, will also enjoy high employability.

Datta's points above are interesting to me in two different ways.

First, look at the second list of 5 items above. He emphasizes (i) innovation followed by *tenacity* in testing, (ii) risk, (iii) risk (!), (iv) risk (!!), and (v) customer feedback for product improvement. This sequence is worth emphasizing: be original and clever, then work hard and long, then watch your risks (repeated!), and listen to customers to constantly improve your game.

Do not think that assessment of risk is restricted to the gas pipeline industry. I have an old friend in the software industry, an entrepreneur, who once asked me the following question.

"I have two options. I could put all my effort into one business idea, with a 35% chance of success. If I succeed, I make X amount of money. Alternatively, I could place my bets on two independent ideas, each with a 25% chance of success, but for each success I would only make $0.3X$ amount of money. Which would you recommend?"

I confidently told him that by the first option he had a chance of making $0.35X$, while with divided bets he stood to make $0.25 \times 0.3X + 0.25 \times 0.3X = 0.15\,X$, so the first option was superior. But he explained that his chance of failure in the first case was 65%, while in the second case it was 0.75×0.75 or about 56%. So, he considered the second option superior because it was less risky. This friend successfully started and sold two companies. Although he still works hard, he no longer needs to.

Let me offer a final story about risk. There is a professor who teaches probability. His erudition is not in question. His child appeared for medical entrance exams, and got selected in a nationally renowned medical college (call it A). The professor told the child to *not* join college A, drop a year, and get into AIIMS the next year. The child did drop a year, and the following year got into neither college A nor AIIMS. The professor made the mistake of thinking that the laws of probability, which apply to the population, would not apply to his child[9].

I will write more about probability and risk in later chapters.

There is another interesting thing about Datta's comments. He is a fully technical person who has done large scale and very practical engineering work. Yet, in his list of desirable qualities in an employee, he does not explicitly talk about technical subjects[10].

To some extent this is because you, as an employee, will learn what the job requires. Your engineering courses teach you how to learn what is needed. A technical education can give you a lot of *transferable* skills, so you are not even restricted to purely technical work.

---

[9] Imagine that your probability of getting into college A is 0.5, and that of getting into AIIMS is 0.2. You appear for the exams, get selected by college A but not by AIIMS. You drop a year and work hard. Your probability of getting into college A rises to 0.6 and of getting into AIIMS rises to 0.3. You have *improved*. But your probability of getting into neither of the two colleges is still $0.4 \times 0.7 = 0.28$, which is a 28 percent chance you will end up much worse than you are now.

[10] Stress, pressure, thermodynamics, fatigue, corrosion ... he had to think about them regularly in his job. Actually, he is an electrical engineer by training. He learnt on the job, and so will you.

CHAPTER 5

Learning English

In a globally connected world, knowing English well is obviously a good thing for your career. If you have read this far, you know some English. However, thinking *about* English may be useful, and perhaps there are some things you want to improve as well.

Let me first discuss language and cognition. This part is strongly influenced by Professor Lera Boroditsky (referred to below as "LB"), who has taught at MIT, Stanford, and San Diego, and whose research is in the fields of language and cognition. I recommend watching her TED talk of 2017 titled *How language shapes the way we think.*

LB reports some remarkable things in her talk.

She speaks of an Aboriginal community in Australia that she worked with. In that community, they have no words for left or right. They use only the cardinal directions: north, south, east and west. As a result, those people have a remarkable sense of geographical direction at all times. There are other consequences. When they depict time, they show it progressing from east to west. So, when they face south, time in their depiction goes from their left to their right; while when they face north, time goes from right to left; and if they face east, then time comes towards them.

LB also says that whereas English has the single word *blue*, Russian has two distinct words: one for light blue (*goluboy*), and one for dark blue (*siniy*). On studying the brains of people who are watching a color changing slowly from light to dark blue, it is found that Russian speakers show a moment of "surprise" when the switch occurs, while English speakers show no such surprise. In other words, language influences our culture and even our brains.

A third example from LB's talk involves bridges. In German, bridges have the feminine gender; while in Spanish, they have the masculine gender. As a result, Germans are more likely to refer to a bridge as beautiful, while Spanish speakers are more likely to describe a bridge as strong.

A final item from LB's talk involves a man accidentally breaking a vase. In English, it is natural for the man to say, "I broke the vase." However, in Spanish, it is more natural to say, "The vase broke." In other words, in Spanish it is less common than in English to lay direct responsibility on the person causing the accident. LB points out that such things may affect how things are remembered, and how responsibility is assigned.

Motivated by LB's talk, in the rest of this chapter I present my views for an individual Indian engineer trying to build a career.

English differs greatly from Indian vernacular languages. English came to us from British rule.

There are some obvious negative consequences, like lower pride in our own languages. But there are also some obvious positives, too, like better international access. Our IT boom, which employed one generation of young Indians, was possible because many Indians knew English.

Less obvious positives are in improved terminology in engineering and technology. Although we do have technical books in vernacular languages, the more advanced ones do not flow naturally.

Let us start with a simple example. There is a Hindi word for the elementary concept of acceleration: *tvaran*. This word is perfectly correct, as far as I know, although usage of the word is not common in daily life outside our technical disciplines. As engineers, we can guess why. Position is easy to see. Speed involves one time derivative, and we can sense it well. Acceleration involves double derivatives, and our natural intuition for it is poor. I think it is not an accident that $F = ma$ came to us from a discoverer of calculus. If typical Hindi-speaking villagers do not know the word *tvaran*, then do they know how to think about acceleration?

As another example, consider efficiency. There is a word for it in Hindi, used in textbooks: *dakshata*. We use it, but it is not exact. An online Sanskrit-to-English dictionary I looked up offered *ability* and *dexterity* as meanings for *dakshata*, which are more related to skill. Friends I have consulted, speakers of Bengali and Telugu, have suggested other words which correspond more closely to capacity, effectiveness, productivity, and the like. If we talk about the efficiency of gas flame cooking versus induction heating cooking, then *dakshata* does not seem appropriate. If indeed there is no precise Indian word for efficiency in common use, then perhaps many Indians who speak only the vernacular languages do not think much about efficiency.

Three whistles!

Over two decades I have interacted with a series of cooks in my home. I have told them that if they let the pressure cooker whistle, the food inside the cooker cools down[11] and the cooking takes longer. Simultaneously, energy is spent in heating and humidifying the kitchen while also making a loud noise. The most efficient cooking occurs when the cooker is maintained on the verge of whistling by lowering the flame. To the last cook, they have preferred to continue with their formulas of "two whistles" or "three whistles." I talk to them of efficiency. But I think they do not completely connect with the idea. I must also admit that perhaps the two-whistle approach has its own advantages. Maybe it ensures fewer errors as my cooks do other things like washing and cutting.

Perhaps efficiency is not crucial for musicians, artists, soldiers, farmers, and cooks. Perhaps they are interested in mastery, beauty, victory, productivity, and reliability, respectively. In contrast, *factories* care about efficiency. Maybe the word became important after the industrial revolution, and we in India never really had one of those.

---

[11] The boiling point of water rises with pressure. When steam exits the cooker through the whistle, the pressure inside drops. This causes some of the water in the cooker to evaporate quickly, absorbing latent heat of evaporation, cooling the food that you are cooking. Meanwhile, the released steam takes latent heat into the kitchen, away from the food.

If most Indians do not have a precise word for efficiency, then daily life should offer many opportunities for savings. I think it does. For example, when we cook rice, we often put a pot with rice and water on the burner, and let the water boil until the rice is done. However, we could presoak the rice in water; and then when the water came to a boil, we could turn off the flame, cover the pot, wrap it in a towel, and set it aside. The rice would get cooked the same, and we would save a few minutes of gas flame. Multiply by a hundred million homes and 365 days per year, because many Indians eat rice every day[12].

Moving up from acceleration and efficiency, even educated speakers of solely Indian languages[13] may struggle with more complex ideas.

For example, what is a word for stress? You may have thought of the Hindi word *tanaav*. But *tanaav* actually means tension, while stress can also involve compression and shear. For engineers, there is a big difference between stress and tension. I have asked some engineer friends who speak other languages for words that mean stress. Some words they suggested are

---

[12] Should something like this not be in our school books?

[13] Assamese, Bengali, Gujarati, Hindi, Kannada, Malayalam, Marathi, Odiya, Punjabi, Tamil, Telugu, Urdu, and many others.

related to pressure, which is the opposite of tension, and also not the same as stress[14].

If the technical ideas we are trying to express are more complex, then the difficulties we face are greater. If you are a student of applied mechanics and want to consider more complex examples, you may like to play the following short game in any vernacular language you like. Play as rapidly as possible, to see where your instincts lie. Explain the difference between the stiffness and the strength of a material, and point out how these two differ from hardness and toughness. Explain that when we bend a rod with pure end moments, plane cross sections remain plane. When we twist a noncircular prismatic body about its axis, plane sections warp due to out-of-plane displacements that are governed by a simple linear partial differential equation. A rotation in three dimensions may seem like it has a magnitude and direction, yet it is not a vector: it is actually a linear transformation. The algebraic eigenvalue problem finds use in solving problems of free vibrations, transient heat flow, structural buckling, and a variety of stability analyses. I think most Indians

---

[14] Actually, one friend told me he saw a Bangladeshi book in which stress is called *peedan* and strain is called *bikriti*. A decent effort! Of course, much remains to be done. Development of an entire language is not the work of one person. Meanwhile, *you* need to build a career.

cannot express such complex technical ideas in Indian vernacular languages. I certainly cannot.

If you want to build a career as an engineer, learn English well. In technical matters, the English language is extremely strong. Language is the vehicle of thought. Native speakers of Spanish, French, Italian, and German, for example, publish many technical research papers in English; but Americans do not publish much technical work in those languages.

Let us move on to some less obvious cultural implications, prompted by the ideas of LB above.

Suppose you send someone to do some particular work for you. In English, you might call that person and ask, "Have you done the work?" And the person will say, "Yes, I have done it," or "No, I have not yet done it." But in Hindi and Bengali we are likely to ask "*Kaam hua?*" or "*Kaaj ta holo?*" And the person will reply yes or no, but in the third person instead of in the first person.

Think about it. "*Kaam abhi tak nahin hua*" or "*Kaaj ekhono hoyni.*" The work is not yet done.

The similarity with LB's observations on English and Spanish, for the accidental breakage of a vase, is striking. Every sentence has a subject, but may or may not have an object. In one form above, the work is the subject, and the actual doer of the work does not appear. In English, the doer is the subject, and the work is the object. Correspondingly in LB's example, in one form, the vase is the subject and the breaker of the vase does not appear. In English, the breaker of the vase is the subject, and the vase is the object. In this way, it may be that Hindi and Bengali are more like Spanish than like English.

The difference between Spanish and English, over many people and many tasks, could have implications[15].

What language means, and how a country or culture should respond to that meaning, is of course a big matter that involves the nation. I am not qualified to comment on such a big thing. I am interested here in something much narrower. I am interested in what language means for *you* as you try to build a career in engineering.

The corporate world as well as the world of engineering, practically speaking, usually pay a premium for good speakers of English. It surely helps that fluent Indian speakers of English can talk to people from many other regions and cultures than their own. The practical conclusion for *you* is clear. Improve your English.

---

[15] It is interesting to compare Spain with the UK. Spain has a significantly lower GDP per capita, and a larger unemployment rate. Has culture played a role? Language affects culture.

Let me now address the process of learning English. Even if you read English well, you may still wish to speak it better. There is a difference. Reading requires recognition alone. Speaking involves other choices: which words to use, how to construct your sentences, and so on.

I offer two suggestions. Both exploit internet resources that are now common.

The first suggestion is to watch programs (cartoons, movies, debates) in English where there are subtitles in English. Watch these programs again and again until you know the entire dialogue. Repetition is key. In this way, you build up a vocabulary of *phrases*. It may be better to choose programs where the dialogue is formal rather than filled with slang. As your English improves, try more complex dialogue. Eventually, consider speeches by excellent speakers.

I know from experience that this strategy works in Hindi. As a Bengali person, I struggle with the gender that is assigned in Hindi to all nouns. In school, one of my classmates once asked the teacher what the gender of *laathi* (stick) is. The teacher was a native speaker of Hindi, but did not answer instantly. He paused for a second, and then said, *"Jiski laathi uski bhains. Streeling."* He who holds the stick owns the buffalo. Feminine. From Bollywood songs, I know *mera dil* and *tera pyaar* (both masculine, regardless of the gender of the singer).

Decades ago, before the internet, a friend of mine improved his English by reading "at least one page" at bedtime. He read many biographies over some years. The strategy worked, and he has had a stellar career, too. However, my plan above may be more fun and take less time.

After you learn the words and phrases, there are still smaller matters within speaking. How you pronounce words, which syllables you stress[16], where you pause in a sentence, and how fast you bring out your sequence of words. Here, too, the internet offers a remarkable opportunity. Listen to the program using headphones on low volume, and *speak* the words as they are spoken. Adjust volumes so that, within your head, the two sound streams (from the program and from your own speaking) are equally loud. Then you will see where your speech patterns differ from those used in the program. Slowly, your speech will change to match the program.

Think of it as a process of smoothening passages and rounding out corners by repeated traversals. If you want more formal theory, you may like to look up the term *neuroplasticity*.

---

[16] Indian languages do not have stressed syllables within words. For example, in the word *develop*, the syllable *vel* is stressed. Many Indians miss this emphasis and pronounce all syllables with nearly uniform stress.

Exercise 5.

It is time, again, for an exercise.

You speak English in a certain way. The aim now is to notice how other people speak.

Think of someone you know who speaks in a different way. Maybe that person is from another state, and maybe their mother tongue is different from yours. Maybe that person is a news reader on TV. It does not matter. Think of one English word which you and that person pronounce in different ways. Pay attention and *notice* the difference in pronunciation. If you can do this, then think of another person, maybe from another state, and a different word. For each such person that you can think of, with a clear assessment of at least one new word that they pronounce quite differently, give yourself one point. One point per person, not one point per word.

Can you score 5 points?

"People speaking English" is not one uniform mass. Different people speak differently. Once you notice how others speak, you will notice more clearly how you yourself speak.

CHAPTER 6

Campus Interviews 1

Let us now think about campus interviews where you are the candidate.

The goal of entry level recruitment by large companies is not really to find the best people. It is to use a quick and reliable method to select several people who are good enough and cheap enough.

Personal interviews are actually not very reliable ways of selecting good employees[17]. However, they are popular because hiring people without even meeting them has other risks.

---

[17] I recommend the book by Daniel Kahneman, *Thinking, Fast and Slow.*

Campus recruitments are preferred over, e.g., walk-in interviews for many entry level positions. This is because an employer can first use a filter to select a "good" college and can also get some decisions out of the way like salary and designation to be offered. The recruiters visiting the campus have no other duties that day, which helps them focus. They look at the pool of candidates in that college and apply some filters. They may apply a grades filter[18], and then a filter of some kind of aptitude test. The final interview may sometimes become merely a test of personal qualities and presentability, provided the candidate does not make any serious technical errors.

Some students think their job search ends when they get their first job. But there is a feedback loop. If someone in your college behaves badly, unethically, or unprofessionally with campus recruiters, then those recruiters may not return to your campus. Word may spread. Some years later you may want to switch jobs, apply to another company, and encounter a person there who is annoyed with your college. That person may discard your CV and forward another one, and you may never find out why.

Let me now offer some advice on the interview itself.

---

[18] If your seniors in college say grades are not important, do not trust them. Grades are a part of the picture.

1. THE POWER BALANCE.

The interviewer, for your first job, is standing at a door and may let you enter the world of employed people. On the other side of the door is, hopefully, about 35 or 40 years of fruitful employment. The interview may last for a few pleasant minutes, but a positive outcome will have a long-term impact on your life. Please show respect for the process, and let your behavior indicate that you take it seriously.

Almost all the power lies with the interviewer. If the interviewer decides not to hire you, there is no court of appeal.

2. MARKET ASPECTS.

In the job market, as in any market, there is a buyer and a seller. You are the seller. The employer is the buyer. If you do not agree, ask: who is paying whom?

You can try to decide if you are in a buyer's market or a seller's market.

If you are from a top college, have excellent credentials, have demonstrable and superior and specific skills that the employer wants, *and* people like you are in short supply, then you are in a seller's market. You can show confidence in your worth, and politely ask if the employer is open to salary negotiation, or volunteer preferences that you might have (e.g., I would prefer

Hyderabad over Pune, or I would like to join project A and not project B, etc.).

Campus interviews and your first job.

In a buyer's market? This means there are few employers, many applicants, and you are one in a big crowd. Big software companies that hire in hundreds from a few colleges, paying moderate but not excellent salaries, are in a buyer's market. The work they offer can be done by many people and you are not special to them. These companies can find other colleges to go to. Here the interviewer may have little patience for your constraints, and be more interested in whether you can be counted as an honest and able contributor. In such situations, I suggest that you should show appreciation for the company's mission, its scale, and its ability to offer you room for professional growth.

If you really have no significant credentials (say, poor grades from a not-well-known college), then you are in a strong buyer's market and perhaps at a disadvantage. But do not give up yet. You could still try to stand out in some way by showing strong ability to contribute. Maybe you are an excellent programmer (say something about it: won prizes in competitions?). Maybe you have exceptional physical stamina, and will work past the point where your colleagues drop from fatigue (won prizes in sports? ran a marathon? are used to working 10 hours in your farm, on hot days, in your village?). Maybe you have excellent people skills (general secretary of your college union? worked for an NGO and spoke with thousands of people?). If you are not exceptional in such ways, you can still try to make employers sense that your commitment and willingness to be a soldier in *their* battle is high. In such situations, your aim is to make them think, "This person will be useful."

3. LYING TO INTERVIEWERS.

I seriously believe you should assume that the interviewer has more experience of interviews than you do. Assume that if you lie, the interviewer will know.

Lies are annoying. I am a teacher and not an interviewer, of course. But as you may guess, students sometimes lie to me. I catch some of them. I am usually annoyed, especially if I have heard those same lies many times since I was a student, some decades ago. The interviewer may feel equally annoyed if you offer easy-to-detect lies.

Unless you are a brilliant liar[19], I think it is best to be honest. The interview room is not a court of law. The interviewer does not have to prove that you are lying. They will simply not select you.

The ability to detect lies runs deep, into past experience and psychology. Let me share a story from my job as a professor. Masters students were presenting their thesis work one afternoon, about 15 minutes each. I was one of a small panel of faculty members attending the presentations and giving marks out of ten. One student presented some equations and then presented a graph. I know something about that graph. The student's demeanor, where he was pointing, the words he was using and the words he was not using, a sudden change in the text style … such things convinced me that he had simply copied that graph from somewhere. I offered him a choice: he could admit that he had copied the graph and accept poor marks, or he could take 30 minutes after his presentation to generate

---

[19] If you *are* a brilliant liar, you may clear the interview but pay for it eventually. It is not easy to repeatedly fool increasingly smart people all the way to the top. Statistically speaking, honesty is very often the best policy. I will write more on this later.

that graph using his own code while I watched. The student thought about it, and admitted that he had copied. Later, a colleague asked me how I had known. All I could say was that there was a cultural inconsistency. Note that as an examiner, I needed to prove fault (which I did by obtaining a confession). As an interviewer, I would simply not have selected him.

I am not special. Please do not underestimate the interviewer.

4. PERSONALITY.

Imagine you have a house with a nice garden. You want a gardener. Some people come and apply for the job. Will you be interested in *their* plans for where they want to plant your flowers on your land? Probably not. It is more likely that you will see whether they are smelling of alcohol in the afternoon, whether they look like they are clean and honest and hardworking, whether they look you straight in the eye without being disrespectful, whether they seem like they will be receptive to your suggestions, and so on.

Now, returning to the campus interview, reverse roles. Talk straight, look clean, show that you are sincere. That is what they want, as you would in their place.

Let me clarify that I am not trying to insult either you or gardeners. The gardener example is meant to help you realize that the psychology of someone who is offering a job is very different from the psychology of someone who is applying for one.

5. HONESTY TO YOURSELF.

Unless you desperately need a job, stay true to yourself and your wishes. Your classmates and you will compare salaries and companies only on the day you receive offers. Then each of you will go off to a company where all fresh recruits in your group are probably paid the same. If you spend each day hating your job, then a few rupees extra at the end of the month may not make up for it. Having a job that you enjoy, colleagues that you get along with, in a location that you like … that is a blessing. Getting a job that directly (even if a bit slowly) leads to wider opportunities and greater scope for contribution is a blessing. Do not walk away from it all just for a few rupees extra, at least without serious thought.

6. THE LONG GAME.

These are tough times. Your generation may see more obsolescence and layoffs than mine did. Even if you are interested in money above all other things[20], note that

---

[20] A one-dimensional approach, it seems to me; but you are the captain of your ship.

your lifetime earnings are not decided by your starting salary. They are decided by your average salary over your working years, times the number of years worked. In case you get laid off at age 50 and cannot find a new job, you will still have living expenses.

Due to inflation, 100 rupees today buy less than they used to buy ten years ago. But we can imagine using inflation-adjusted rupees, and do a simplistic calculation to see something basic.

Let us say, in job 1, your average lifetime salary is 10 lakh rupees a year, and you work from age 22 to 62. Then you retire, and live until age 82. Perhaps you get used to living on half your salary, in which case you earn 4 crore rupees (10 lakhs per year times 40 years) and spend 3 crore rupees (5 lakhs per year times 60 years), reaching the end of your life with 1 crore rupees which you leave to your children. A financially unexciting, but satisfactory, life.

Compare with job 2, in which your salary is 20 lakh rupees a year, and you work from 22 to 52; then you are laid off, cannot find other work, and again live to the age of 82. Perhaps you again get used to living on half your salary, in which case you earn 6 crore rupees (20 lakhs per year times 30 years) and spend 6 crore rupees (10 lakhs per year times 60 years), dying with nothing left. Aging toward death with no money in the bank cannot be fun. Add to this the frustration of having no work to define yourself while you still have working years left.

If you get two job offers, I suggest you pay less attention to the starting salary and more attention to which job will keep you employable in a sustained way. Which company is such that, 20 years later, the chances are better that you could walk out and easily get another job? Recall Exercise 2 (Chapter 2).

Exercise 6.

It is time for an exercise.

I gave a list of 6 items above. Before you look through the following paragraph, glance back at the list of 6 and see if you agree with the sequence in which they are presented. See if you can guess and represent the dominant principle with a few words in each case. Now see if your take-aways match my intended takeaways below.

(i) Respect the process, (ii) Know where you stand, (iii) Be honest to them, (iv) Be ready to contribute, (v) Be honest to yourself, and (vi) Take a long view.

This super-shortened list of 6 items is not actionable. It is a memory aid, not a plan.

Chapter 7

Annual Coin Tossing

Before you start reading this chapter, recall Exercise 1 from the end of Chapter 1.

Many Indian college graduates, including engineers, appear for the UPSC exams. Success can lead to prestigious jobs with the central government, with job security, respect in society, influence, and even a chance to make a positive difference to the country. The dream is powerful. But success rates are low. Many candidates appear several times in a row, spending a few years attending coaching classes full time. Most of these candidates eventually fail to get selected and go on to do other things.

This chapter contains advice to a student who gave a few years to preparation for the UPSC exam, but failed, and now wants to pick up the pieces and move on.

Imagine tossing a coin to determine your career. Imagine that the spinning coin flies very high up, so high that it takes a whole year to return and land. When the coin does land, the chance of a win is low. Maybe one in ten, or maybe one in a hundred, but typically much lower. The attraction of the game is that *if you win*, you win big. Would you play this game? The UPSC exams resemble this game.

You may think my coin toss analogy is wrong, because (a) the candidate actively prepares for the exam according to a definite plan, and (b) the coin toss is fully random with no place for preparation. But the analogy has merit. Although success for some candidates may be more likely than it is for others, nothing is guaranteed in advance. The candidate's performance on the exam and the outcome of the coin toss are both impossible to know in advance. Both are known only afterwards. The relevant technical term is *random experiment* (look it up).

For years, I had not thought about these candidates much. I just thought they were chasing a dream outside engineering. Then I found the following question on *Quora*.

I failed UPSC 5 times in a row. Now I am frustrated and unemployed at the age of 27. I never had a job. I graduated at 21 as a mechanical engineer. Sometimes I feel worthless. What should I do?

What follows is adapted from the answer I wrote there.

If you have given some years to failed UPSC attempts, then you are not alone. The success rate on this exam is about 0.1%. About 1000 succeed, from more than 1000000. The vast majority of aspirants do not get in.

Individuals make up the population, but each individual thinks that the statistics of the population will not apply to them.

Suppose you are in the top 1% of the population. That means, out of every random 100 people, most of them working hard, all dreaming of the UPSC, most of them taking coaching, you are comparable to their top candidate. You are the best in 100. That is not bad at all. Think of your high school, and imagine being the strongest or most good-looking person in a class of 100 students.

For the UPSC, however, being in the top 1% means your chance of success is 10%. Or your chance of failure is 90%. If you appear 5 times with a 10% probability of success, the probability of failure is 0.9^5 or about 60%.

Some may quibble about the assumptions behind my calculation, but I hope the principle is clear[21].

Look hard at this number. Your emotions will try to stop you from seeing it for what it is. If you are a top 1% candidate competing for a 0.1% slot, then your probability of success after one attempt is something like 0.1; and if you try five times your probability of success is still something like 0.4. An added problem is that, while you *think* you are in the top 1%, you may actually be in the top 2%, making your chances poorer. I read somewhere that in a large study, 90% of people polled believed that they were above average.

Voluntary writing on public fora suffers from survivor bias. People who succeed at something difficult tend to write about it and claim that it is easy. People who fail mostly stay quiet. Casual readers think there is a winning strategy for the UPSC exams. There cannot be. If everyone uses that winning strategy, the success rate will remain 0.1%. Casual readers who read the previous two lines will think that they will find a winning strategy which *others do not know*. But there are lakhs of people trying, and thousands of very smart

---

[21] I am not sure this trivial idea bears fancier calculations. But if you have taken the exam five times and failed, then you are more likely to have been in the bottom half of the top 1% rather than the top half of it, in which case your true probability of success after five attempts may have been poorer than 0.4. We will never know.

people running coaching centers whose business is to think of good strategies. Your chances of finding a winning strategy by yourself are tiny. If you like such small chances then, as you walk on the street, perhaps every ten steps or so, bend down to pick up whatever is lying there to see if it is gold.

The way to success, for the population, does not lie in coaching classes. When everybody takes professional coaching, it is true that the individual must take coaching or their chances are poorer. But if everyone takes coaching, the overall success rate after coaching is still 0.1%. Coaching classes do not sell success because there is not enough success to sell. They sell *hope*, which is limitless. In a country where good outcomes are limited, hope sells briskly. The main thing achieved is that the coaching institutes get rich, because the person who does not take coaching is almost guaranteed to lose[22].

You will have heard of one or two people who got in without coaching. That is survivor bias. You may hope that the same thing will happen to you. That is the individual's hope that the population statistics will not apply to him or her. Recognize your emotion-laden thinking process for what it is. Probability has no mercy.

---

[22] As of September 1, 2023, the popular press says there have been 23 suicides so far among coaching center students in Kota this year.

It may help to clarify your thinking by considering an excellent way to do public service: voluntary blood donation. A healthy adult can safely donate slightly more than 1 liter of blood over one year (on three to four separate occasions). Now imagine a hypothetical situation where the government decides to reward UPSC candidates who do public service in this way. Imagine that any candidate who has donated 1 liter of blood in the preceding year will be given 5 marks extra on the UPSC exam. How many liters of extra blood will be donated each year? The number may be quite large. How many people will qualify for the UPSC? The *same number*. Why is this hypothetical example important? Because coaching classes cannot take your blood. But they can take your energy, your money, and your youth.

It may help to clarify your thinking further if you return to the coin toss game described at the start of the chapter, and realize that you have been playing it for five years. Because you were studying, you forgot about the randomness. But all candidates were studying, and the randomness remained. Many lakhs of people play this game every year. Most of them lose.

So, you played and lost. Do not lose heart. You are still young. Many roads still lie ahead of you. I merely suggest that, having taken this hit, you do not immediately try other low probability competitive events.

If you are from a business family, join a relative's or a friend's business. I suggest that you should not start a business right now.

If you are from an academic family, consider higher studies. Since you have studied engineering, I can offer something specific. If you are from a centrally funded technical institute (CFTI) and your undergraduate GPA was more than 7.5, you will qualify for direct PhD interviews in the IIT system (that rule exists at the time of writing this book). You may start at age 28 and finish by the time you are 33, which is not terrible for a PhD. The PhD erases earlier gaps to some extent. But the career track is different and you have to be ready for that change (more on this topic in a later chapter). If you are not from a CFTI, you may consider the GATE. Since you are used to competitive exams, maybe you will do well. With a high GATE score you can join an MTech program, or even direct PhD programs. Some foreign universities (in Germany and Singapore) accept GATE scores as well. If your English is very good, you can consider the GRE or the GMAT, and seek higher studies abroad. Finally, you can appear for the CAT and study for an MBA.

If you have extended family members or friends who work in engineering companies, you can send out feelers to them and see if someone can help you get a job. This is better than blindly uploading your CV on job portals. Understand that they are doing you a favor. Let me share a story here. I know two men who are cousins, one a few years older than the other. Several years ago, neither of the two had a job. The older one found some funds and started a medical testing lab. The younger one requested an opportunity there, and was offered some work that he considered to be of a low level. He turned down the offered work. As it happens, the lab idea did not work out. Eventually, the older one found a good job in an upscale store chain that sells shoes, clothes, and such items. He began to do well. The younger one again asked for help. The older one mentioned that anything he could arrange would be lower than his own position. The younger cousin, offended again, declined. When I last heard, the older one was doing well. The younger one, having done some BPO work for a while, was unemployed and did not seem to be looking hard any more. Reality is cruel. Help in getting a job is a huge favor. Be humble, and be grateful. Then build a career. Then help somebody else.

If you cannot find a job through such contacts, do not give up.

If you are actually good at multiple choice question (MCQ) type exams, then you might get a job in a coaching center as well. While I do not really think such jobs are very interesting, they can pay well.

Search through the networks of friends you may have made while you were preparing for the UPSC. Many of them may also have failed to get in. What are they doing? Maybe they can help you find work.

Whatever you do, remember that the population statistics will probably apply to you. If you are optimistic, hope for the mean plus half a standard deviation. If you are pessimistic, be prepared for the mean minus one standard deviation.

Watch out for people who say something is easy. They are probably lying, maybe unintentionally. In India today, career-wise, nothing is both excellent and easy. But labor, persistence, and moderate luck can lead you to moderately good outcomes.

Let me close with something about psychology.

You have three options. (1) Earn enough to support yourself and your family, (2) be dependent on somebody else, and (3) starve. Of these three, option (1) is best. To achieve that option, you now must take control of your life. Control exerts a psychological burden. Hope is light and sweet.

One way to achieve option (1) is to clear some competitive exam. This is what you have tried. When you prepare for such an exam, you work hard; but control remains with the exam authorities so your psychological burden is light. Day to day, you have hope, which helps you to spend another year.

But a time comes when most people have to admit that they will probably not clear the exam. At that time, they can continue to live on hope, refusing to take control. By this time, hope is just a drug and they are addicts. Postponement of taking control is making them weaker, not stronger.

At some point, most competitive exam aspirants have to either take control by doing something more directly aimed at option (1), or stay hooked to hope for another year. The problem is that time does not stop, the battle does not go away, and each year the positive kick from the hope-drug gets weaker and weaker.

I have painted a somewhat gloomy picture above, but the numbers do not allow otherwise. If 1 in 1000 will get in, then 999 will not, and there is nothing anybody can do about it.

You must take control. You do not really have a choice.

Exercise 7.

Let us end with an exercise. In this exercise you will think like a businessman.

I will use hypothetical numbers below. Replace them with your own numbers if you like.

Suppose there is a tough annual competitive exam with a population success rate of 0.1%. Nationally, 1,000,000 students appear for this exam each year. You own a coaching business and you train candidates for this exam. You have 10,000 enrolled students. You are a small player, but you are ambitious. You want to be a big player and earn many crores of rupees.

You have 50 teachers, i.e., one teacher for every 200 students. Your coaching quality is very good. Your success rate among your students is 1%, which is ten times bigger than the population success rate. This means 100 of your students succeed every year.

You charge your students 50,000 rupees each in annual fees. You pay your teachers 20 lakh rupees each per year. You also have 50 classrooms, and you incur various costs totaling 50 lakh rupees per classroom per year.

So, your annual fee income is 50 crores. You pay 10 crores in salary and 25 crores in other expenses. That leaves you with an annual profit of 15 crores. You want more.

Suppose you work harder and your success rate increases by 1%, that is, instead of 100 people succeeding each year, you now have 101 people succeeding. Your profit changes by zero because enrollments did not increase. You learn that you need to advertise more than you need to improve your success rate.

Suppose you spend 1 crore on advertising, telling people about your success stories with photographs of smiling "winner" students. This causes your enrollments to increase by 4%, i.e., you go from 10,000 students to 10,400 students. You still have the same 50 teachers, only there are now 208 students in each classroom. Instead of individual chairs you use long benches so that they squeeze in and fit. Your cost per room does not increase. Your income increases by 2 crores, so after subtracting the advertising expense of 1 crore you have increased your profit by 1 crore. You want more.

You have an idea.

Your 50 teachers are not all the same. Some are better than others. You give internal exams to your students and select the 1000 top candidates from your pool of 10,000 students. These 1000 students are assigned to your 5 best teachers. So now you have 5 good classrooms and 45 mainly money-earning classrooms. Matching good students with good teachers increases your success rate, and you get 110 successes. You spend 1 crore on advertising, and this time your enrollment increases by 10%. Your income increases by 5 crores, so after subtracting the 1 crore spent on advertising you have increased your profit by 4 crores. That is better.

From these experiences you conclude that your profit lies in maximizing enrollments, keeping each classroom as full as possible, matching the good students with the good teachers to increase your overall success rate (although this means the weaker students have even lower chances of success), and advertising, and advertising, and advertising. You realize why national newspapers have full front-page ads from coaching classes day after day.

One day, a young student from a village comes to you. He is weak both academically and financially. He got poor marks in college, and his family sold land to pay for coaching. You know his chances of success are negligible. If you accept him, he will be sent to one of your 45 money-earning classrooms and not one of your 5 good classrooms.

Will you offer him a cold drink and a chair under the fan, smile at him, and admit him? Or will you emphasize to him that his chances are negligible?

Are you familiar with the saying which goes: if the horse makes friends with the grass, what will he eat?

CHAPTER 8

Long Term Value

This is the chapter with my main list of action items for engineering students. Some items are obvious, and I have discussed them briefly. Others took me several years to appreciate, and I have given them more space.

1. LEARN TO SPEAK.

Speaking good English will get you an advantage in India, as I have discussed already. Even if your English is not excellent, avoid obvious grammatical errors if you can. Even if you speak only in other languages, speak logically and clearly.

Your speech must be clear. Do not mumble. Let the last word in the sentence be as audible as the first word. In a meeting, the closest person in the room should not feel that you are shouting. The farthest person should be able to hear you.

Do not answer a question before it is finished (listen). Do not answer a different question (be aware of what is being discussed, and stay on track).

Some examples may help.

Question: "Did you take a bus?" **Answer:** "Yes" or "No" or "No, I walked" or "No, I took a taxi."
Answer to a different question: "I live far from this place", or "I got delayed by the rain."

Question: "What was your BTech project on?"
Answer: "It was on experimental strength characterization of thin sheets of a new high strength titanium alloy."
Answer to a different question: "Titanium alloys are very good" or "It was on a universal testing machine."

Learn how to give a good technical seminar where you have a microphone, an audience, and twenty minutes. I have written about this earlier.

Learn how to adjust the length of your presentation based on who is listening and how much time they have.

2.   LEARN TO WRITE.

You should be able to write in simple, clear English. This is more than learning grammar.

For example, take a 1000-word essay. Choose what you will keep and what you will discard and how you will reword things, in order to rewrite it in 800 words. Get feedback from someone on whether you did a good job. Then make it 600 words. Get feedback again. Then 400 words. Then 200 words.

Read technical articles and good textbooks. Study how the paragraphs are structured. When does one paragraph end and the next one begin? What does one paragraph contain? Study the sequence and the flow of ideas. Compare them with technical reports that you may be writing.

Most people do not know how to write satisfactorily. The problem is very big. For example, a Google search for

"how to write a research paper"

offers millions of results. If it was easy, would so many people feel compelled to discuss it? Careful writing will sharpen your reasoning, too. Each helps the other. Learning to write is hard work, but worth it.

3. LEARN TO DO THINGS.

When you go for a job interview, you are a person and not a grade card. Develop some skills and activities outside your syllabus and classroom to help define the *person* better.

Here are some examples of things you might do. Learn welding, or electronic gadget repair. Spend some time as an assistant to a motorcycle mechanic. Make an optical handheld device that can measure the speed of a ceiling fan. Run a marathon. Volunteer with an NGO and go to villages and figure out ways to increase cooking or lighting or water usage efficiencies. Volunteer in a community cooking event during a festival and work until you are exhausted.

Such experiences will give you confidence as well as something to talk about.

4. LEARN TO THINK.

Your thinking ability will stand by you. It will set you apart from others.

It is not easy to learn to think. One does it indirectly.

Think up new problems that have not been assigned to you. These problems can be related to your syllabus, slightly harder than what you are used to, and made up by you. When you start asking questions instead of just answering them, it will change the way you think.

Assess problems, don't just solve them. What is asked? Are there multiple ways to approach a given problem? If the problem is changed in this way, or that way, will it remain solvable? Will it become much harder or much easier?

If you have a teacher whose thinking and presentation seem clear to you, or a person on YouTube whose technical presentations seem very clear to you, listen to them with great care and see how their thinking differs from the way in which you might approach the same issue. Would you have arranged and presented your ideas in the same sequence?

Tutor students who are junior to you or weaker than you. Try to understand how they think.

5. LEARN TO LEARN.

Very practical inputs on learning college lecture material can be found on the internet. Look for *Cornell Notes*.

But I mean something more general that will work as you try to develop your understanding of new subjects. Think hard about how learning happens with you. Build up your ability to learn things.

In engineering you have an advantage, in that many topics are taught with some practical angle, and with practical ways of simplifying the analysis. You can try to understand what makes a book *good*. Sit with a few good textbooks on different topics open before you, on a lazy afternoon, and try to find a way of describing the approach of these books that is not tied to the specific subject. What is done again and again?

Problem area definition, basic principles, specific problem definition, basic observations, simplifications, further observations, governing equations, analysis, results, interpretation.

You will find some subjects easier than others. In college, my friend Ashutosh was better than I was at mathematics. I did better at the strength of materials. A few minutes before an exam in mathematics, Ashutosh called me over and said he thought a certain topic might be on the exam. He opened a book and we hurriedly looked through a solved problem. Then we went for the exam. Ashutosh was right! That problem was on the exam. But I could not solve it. I cannot learn mathematics in such haste. After the exam, when Ashutosh heard that I had not solved the problem, he threw up his hands in shock. "But we *saw* it just before we went!" He kept shaking his head for some minutes, unable to believe it. Later, for the strength of materials, I was comforted to find him staring at a book on his table, his wrists moving slowly in the air as he tried to understand some problem with simultaneous bending and twisting of a rod. Our brains were different.

If you find a subject difficult to learn, give it the time it needs. Make sure you are rested. Do not jump from book to book, or website to website. Choose one or two good ones, and give them the sustained attention they require. It is not the *book* that requires time. The subject does.

Some subjects are hard. What is hard can often be handled with patience and a plan. If that fails, try a broad-brush map first, learn the main ideas, and fill in details later. If that fails, try to master *one* idea first. If that fails, find a problem that is based on that one idea, and study the solution. If you cannot understand that solution, see if you can understand the solution to a simpler problem. If that fails, try studying with a friend. If *that* fails, rest for a while and fight another day.

And if you still cannot learn it, learn something else. Life is short. There are many things to learn, and neither you nor I will learn them all.

6.   LEARN TO RELATE.

Throughout your studies, see how and where the things you are learning are applicable in the world outside. In the Indian educational system, there seems to be a wall between the classroom and everything outside it. Your aim is to break your personal wall.

Here are some random suggestions.

When you study mechanics, perhaps you can think about stresses in your bicycle, the pulling force needed to dislodge a nail from a wall, the bending in your shoes.

When you study electrical machines, think about how you might measure the actual power consumption of your ceiling fan. Study the overhead wires which supply electricity to trains. How close can you bring an electrified wire to a ground wire before sparking occurs?

What force is needed to crush a sugar crystal between spoons? What moment is needed to break a pencil? Why and in what direction does wood split?

What force is needed to lift the weight on a pressure cooker to release steam? Can you estimate the pressure? From steam tables, can you estimate the cooking temperature? Is the estimate reasonable?

Your engineering subjects should not stay inside your books. Bring them outside into your daily life.

This will affect your interview performance, your performance on the job, and your ability to do new practical things for a startup. It will affect the things you ask if you visit a factory.

7. LEARN TO FORMULATE.

The main sequence of engineering analysis is simplification, problem statement, mathematization, solution, and interpretation. Practice with problems that allow themselves to be treated in this way, and see if you can reduce them to models or equations that you know are tractable.

This kind of work is genuinely difficult. If you are alert and persistent, you can improve over years. It helps to start early, when you are still young and your mind is agile.

Sometimes such work is called modeling. Sometimes it connects with approximate analysis, or with quick design. If you watch lectures on applied engineering topics from other fields than your own, you may find new examples of such thinking. Sometimes a rule of thumb used by practitioners ("we like this ratio to remain below 4") may not have an obvious basis, and can serve as a starting point for your thinking.

This is a background mental game you can play for years. Get an idea from external interaction, think about it alone until you are stuck, then ask someone or read some more, and then think by yourself again. The goal is not to reach somewhere, but to learn to move alone. Personal growth is a solitary activity.

It is not a race. Give it time. Broad skill in translating real-life problems into solvable mathematical problems is valuable *because* it is hard won.

8.   LEARN TO SOLVE EQUATIONS.

Use a computer to solve equations. Matrices, differential equations, systems of algebraic equations. This will follow the previous step. I find that using a computer to solve a variety of small equations improves my understanding of both the application areas as well as the mathematical background.

In this, I like the software packages Matlab and Maple. Perhaps you can get access to them, or use equivalent packages.

When you solve an equation, do not think of it as a process limited to going from problem assigned to marks received. Get out of that self-limiting tunnel mode and look around.

For example, you may notice that the equation

$$x + \sqrt{x} = 3$$

can be solved as a quadratic equation for $\sqrt{x}$. But do not stop there. The equation is not an isolated thing. For example, it is obviously not far from

$$x^{3/2} + x^{1/2} = 3,$$

which can be solved as a cubic polynomial; and also not far from

$$\frac{dx}{dt} + \sqrt{x} = 3,$$

which requires different thinking. What about the system

$$x + \sqrt{y} = 3, \quad y + \sqrt{x} = 2?$$

The same principle applies to bigger and harder equations. When you solve one equation, you develop skills that can help with other, nearby, equations. Look around a little, and study the neighborhood.

9. LEARN TO PROGRAM.

You should know one or more computer languages, several numerical methods, and perhaps some commonly used software packages. Basic engineering literacy today includes the ability to do some level of computer programming.

10. LEARN STATISTICS.

To look at data, you need to understand the basic principles of statistics. I do not mean things like just the mean, median, variance, and linear regression or hypothesis testing. These are important, but not enough to change your thinking.

Your thinking changes when you realize how limited data sets can suggest things that seem obvious, or natural, or intuitive – but are wrong. Your thinking changes when you develop ways of assessing what the data actually says, and which part could be random noise, and how a rational defense of your conclusions can be started.

For example, consider the following data points:

$$-0.30, 0.02, 0.17, 0.22, 0.24, 0.32, 0.37.$$

You might need to decide whether these data points can be treated as normally distributed. It turns out there is something called the Kolmogorov-Smirnov test for normality, and this data passes the test (by which we really mean that it fails to fail the test).

So, is the distribution normal or not?

To make the dilemma clear, imagine that our data points were instead the cubes of the original data, i.e., $-0.30^3, 0.02^3$, etc.

As far as the Kolmogorov-Smirnov test can detect, this new data set is also not significantly different from normal. Yet, the theory is clear: the cube of a normally distributed variable is not normally distributed.

So, returning to the first data set, can we treat it as normal or not? Imagine that the data consist of measurements in your factory, and much depends on the conclusions you draw from them. As an engineer, you have to decide, and your decision has consequences.

You may ask, why not just get more data? If these are ball bearing diameter error measurements in millimeters, taken from a production line, then perhaps we can get more data. But if these are data from machine breakdowns that happened randomly and cost the company a lot of money every time, then this may be all we have. And on the basis of this data, you may need to take an important decision: whether to continue running night shifts, whether to buy a more expensive machine instead of a cheaper one, or whether to lay off a person who seems to cause too many accidents.

Statistics is a powerful tool, when used wisely, for taking reasonable decisions based on limited data. A nuanced understanding of statistics is of great value. It takes time to acquire, because our brains are wired for individual causal stories and not statistics.

11. LEARN TO USE SOFTWARE TOOLS.

Software packages, and their use, should be in your CV. That is modern life. Packages for engineering analysis, statistical analysis, neural networks – see what you can find and master.

Software tools tend to have a finite time frame of utility. Applications change, computers become more powerful, and software is upgraded. From the viewpoint of long-term value, one must remain current with software tools relevant to one's work.

Exercise 8.

It is time for an exercise.

The main message of this chapter is to select useful attributes, work hard, and improve those attributes. The background message is that if you rise higher in useful skill sets, then future job applicants cannot easily take away your job. In this exercise, we will try to consolidate this idea. You will need a pencil.

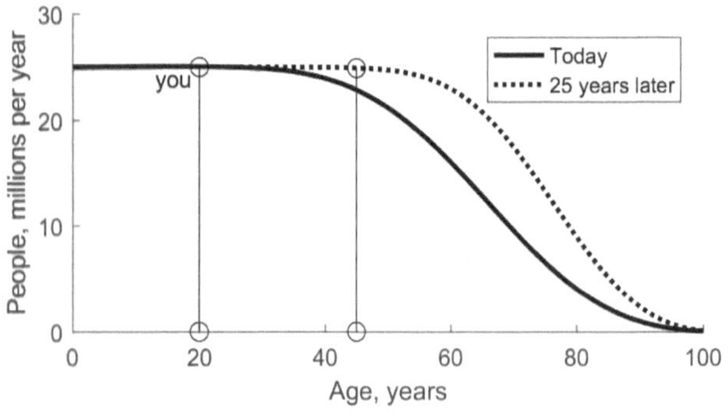

Schematic depiction of Indian demographics, extrapolated.

As discussed already, at younger ages, the population density of India is 25 million per year. In the figure above, population density is plotted along the vertical axis. On the horizontal axis is age in years. The area under the curve represents total millions of people.

The solid curve depicts (approximately) the population of India today. *You* are represented using a small circle on the solid curve, at age 20. Move forward in time to 25 years later, and these young people will be older and new ones will have been born. This will give a population approximately like the dashed curve. You will move horizontally, 25 years to the right, and be 45 years old, at the place marked with another small circle. Note that there are actually four small circles in the figure, and they define a rectangle.

Shade this rectangle with a pencil.

This rectangle represents 625 million people, between ages 20 and 45. These people will be younger than you, behind you, interested in possibly taking away your job. These people will all have had mobile phones and internet connections, and could in principle have learned useful things online.

That rectangle you just shaded is my primary reason for urging you to develop long term value.

CHAPTER 9

Mathematics

A usefully mathematical mind has great long-term value. Conversely, not being comfortable with mathematics will limit the things that you can do. This chapter presents a somewhat abstract model for how mathematics enables and enriches us.

Many people believe that (1) mathematics is a *tool*, and our goal is to understand the physics, and (2) when we do mathematics, we should keep track of the *physical interpretation* of our calculations. I used to be one of those people, and even prided myself on some of my physical interpretations of mathematical calculations.

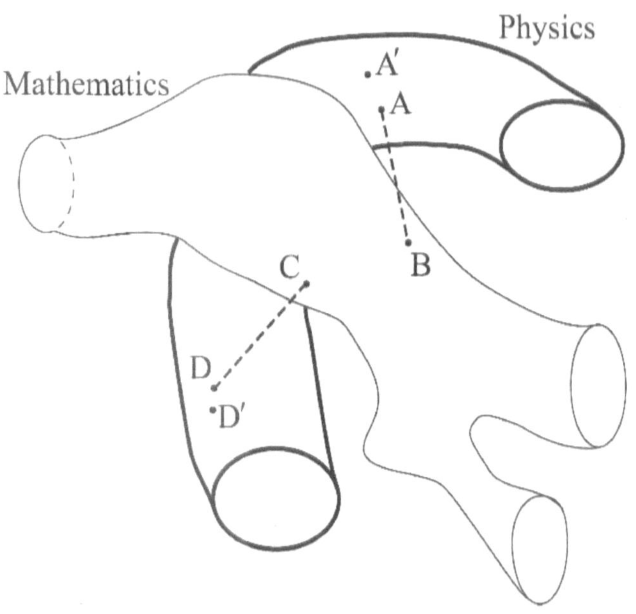

Physics and mathematics as separate but intertwined worlds.

I take a more symmetrical view now. Both physics and mathematics play important roles in engineering. I think that mathematics and physics can be viewed abstractly as intertwined but distinct worlds with occasional bridges or connecting passages. The cartoon depicts the idea. The physical world is depicted abstractly using a thick line, and mathematics is depicted abstractly using a thin line. Some bridges have been found, and we can mentally cross over from physics to mathematics and back, e.g., between points A and B, and between C and D.

Crossing is made worthwhile by two things. Let me first state it abstractly, and then give a specific example. In the physical world we are interested in a real thing A', and want to make some kind of transformation or journey to another real thing D'. Luckily, A' is close to some point A that is simple, idealized, and easy to mathematize. Similarly, D' is close to some point D that is easy to mathematize. In the physical world, a journey from A' to D', or even from A to D, may be difficult or impossible. The journey in the mathematical world, from B to C, is easy, or at least possible. And this is why engineers do mathematics.

Let me offer an example. Suppose I have the design of a shell in mind. That is point A. It includes material specifications, dimensions, support conditions, and so on. This design is precise. The actual shell built will have imperfections, and will be a less precisely known physical quantity A'. The mathematical model of the shell is B. I compute the pressure at which shell B will buckle. That computed pressure is a number, C. It corresponds to a physical pressure D. The actual buckling pressure for shell A' will be some nearby pressure D'. If we did not analyze the mathematical model and built and physically tested the shell, then we would be going from A' to D' completely within the physical world. The journey from B to C may require a study of the theory of elasticity, structural stability, partial differential equations, the finite element method, and eigenvalue problems.

The physical interpretation of a mathematical calculation, if we have such an interpretation, is welcome but not essential. In fact, the greatest benefits from knowing mathematics can occur when the distance between points B and C in the mathematical world is *zero*. In such cases the advantage of using mathematics is infinite, and physical interpretation is impossible.

My favorite example to students, of such a remarkable collapse, is the Laplace equation. This equation governs, among other things, the shapes of nearly flat soap films and also purely in-plane heat conduction in thin sheets. The two physical problems *seem* to have nothing in common. But, to one who knows the crossing points, they do. One can in principle use a wire loop and soapy water to calculate temperatures in a plate without doing any mathematical work. Is this physics, mathematics, or magic? You decide.

To thrive on such connections in a broad way, one must have comfort with many crossing points, with the process of crossing back and forth, and even with finding new and useful crossing points on one's own. A skill of this kind has use both inside and outside engineering.

The two-worlds analogy is more general and useful than the figure above initially suggests. Even within mathematics, we may think of separate worlds of formulas and figures. Some problems involving equations can be simplified if we draw a graph. Some problems of geometry can be simplified if we use equations with vector notation. Students wishing to develop strength may first like to solve the problem using the easier approach (formulas or figures), and then see if the alternative approach could be used (figures or formulas, respectively).

Someone who is comfortable with crossing back and forth between physics and mathematics, has done it for many problems, and can do it for new problems without getting lost - such a person is valuable for the long term. It is not an easy skill to develop, I agree. It takes time, good examples, and much practice. But it is worth the effort. Recall the "learn to formulate" suggestion made earlier in the book.

Do not learn mathematics as a list of problems prescribed within narrow boundaries based on coursework. That way lies clerical aptitude. I offer the two-worlds-and-crossings model as a possible alternative guiding principle.

As a graduating engineer, your career may go towards design, manufacturing, finance, consulting, marketing, data analytics, or other paths we cannot now foresee. The physical world of the figure may then be replaced with a world of lived experience. The mathematical world may then be replaced with one of idealization, representation and manipulation. Knowing both worlds, and being skilled at crossing back and forth without losing your way, will give you an advantage. I believe that students of traditional engineering subjects can be good at this kind of crossing back and forth.

Exercise 9.

A simple exercise.

Recall, or read a derivation of, the governing equation for nearly flat soap films and also the equation for purely in-plane heat conduction in thin sheets. Satisfy yourself that the Laplace equation applies to both, although the physics is quite different.

Recall that, or read books to see how, eigenvalue problems govern both bucking and vibrations, although the physics is different.

Will this exercise help you to embrace abstraction in mathematics?

CHAPTER 10

Campus Interviews 2

As your career progresses you may eventually be an interviewer and not a candidate.

Interviews are often unreliable ways of choosing good candidates. Kahneman has written about this in *Thinking, Fast and Slow*, which is a book I recommended earlier as well. Yet, job interviews seem unavoidable, and you will have to think about them.

Some of the risk in interviewing is reduced by prior use of heuristics. You go to a college you trust. You set a grades cutoff, and perhaps use a written exam to screen candidates further. You put in some interview marks for having a professional appearance and for speaking clearly, because these are important on the job. Only then do you interview candidates, ask them questions, and evaluate their answers.

There are some popular questions that do the rounds. Some of those questions are not good. "Why are you interested in working for our company?" is one of them. Another one is, "Can you sell me this pencil?"

When a question is asked by one person and answered by another, there should be a mutually understood and clear framework using which the correctness or quality of the answer will be judged. Both questions above fail this elementary test.

Look at the first question (why our company). The question is meaningful if someone is considering leaving one company and joining a completely different company (e.g., moving from a big company to a startup). However, that does not apply to campus interviews. For campus interviews, here are two possible honest answers:

a) Well, ours is a small college, and only four companies have come to our campus. I interviewed with the first two and have not been selected. You are third. Tomorrow I will try the fourth one.

b) I studied engineering for four years with the intention of looking for a job. And now, here you are. (Smile.)

Either answer will probably get the candidate thrown out. The interviewer does not *want* the truth, which is mundane and possibly depressing.

This question is merely a challenge to guess a good lie that will tickle the interviewer. The question is worse than "Guess my favorite color," where at least the pointless guessing game is out in the open. The why-our-company question is particularly bad because there are many discussion forums where people offer various engaging, surprising, or cute answers to the question; and you, the interviewer, have *not* seen them all. So, if you get an answer that pleases you, you have no idea (i) whether the answer is original, and (ii) whether the cuteness of the answer translates to value on the job.

The why-our-company question merely says the following to the candidate:

You do not have a job and you want one. I have a job and I could possibly give you one. So, young job hunter, please massage my ego for two minutes.

It makes no sense to waste interview minutes asking a question that (i) has no correct answer, (ii) cannot tolerate the truth, and (iii) has no discernment along the direction in which you are trying to measure something, namely the candidate's worth. Perhaps a more useful question is, "As you know, our company does such and such. Have you taken courses or done projects that are related to this kind of work? Please tell me about them." Give the candidate a framework. Observe how he or she organizes the facts and makes a case.

Next, consider the pencil selling question, which is plain bullying. Some might say that candidates might as well get used to it because the world is a hard place. But you are not there to prepare the candidate for the world. You are there to select a good candidate for your company. This question does not serve your task.

Sell me my pencil.

The candidate could, for example, take the pencil, point out its excellent writing qualities and the comfort of its grip, say something technical about how marvelous the lead is, and mention that the pencil is inexpensive. And an interviewer could claim that the candidate should have discussed volume discounts, or logistics, or wood grain, or fashion, or psychological angles. Or, failing that, the interviewer could just shake his or her head and say, "Sorry, not convincing." Or the interviewer could decide the answer was excellent without knowing if the candidate was trained to give such answers on some internet forum.

There is another type of question which is dangerous. And that is a question which has a simple, straight, unique, factual answer, but you do not know what knowing the answer means.

What is the density of steel? 7800 kg/m^3, as it happens. If the candidate is a metallurgist applying to a steel company, this might be considered basic. Not knowing is unacceptable, and yet asking the question is a bit of an insult. Ask the question if you must, though you should have other things to ask as well. But if the candidate is an economist or an electrical engineer, you should know in advance what the implications of a correct or incorrect answer are.

Why are manhole covers round? The answer is rather pleasing: noncircular shapes can fall into the hole if they are rotated and tilted, but circular shapes cannot. Unfortunately, the world has two kinds of people who can answer this question. A tiny group that can figure this out within a minute of thought, and a much larger group that has already read this somewhere. A random candidate who answers the question is highly likely to be in the second group – and how would that help your company?

I suggest that you think of each question as the beginning of a conversation. You, the interviewer, have experience. You know the likely directions in which the answer could go. For each direction, you have some plan for guiding the conversation in ways you like, as you watch the candidate think.

My own main interviewing experience is for admission to postgraduate programs, where the style differs a little but perhaps the principles are the same. Here are some examples (almost real, only slightly edited).

1. Me: Please solve $x + x^{1/2} = 2$.

Candidate: $x = 1$.

Me: Not bad. You got that from observation. What would you do for $x + x^{1/2} = 3$?

Candidate: I am not sure how to proceed. A numerical solution?

Me: What is a numerical method that you like?

Candidate: The Newton-Raphson method.

Me: Tell me about the method.

etc.

2. Me: Please solve $x + x^{1/2} = 2$.

Candidate: $x = 1$.

Me: Not bad. You got that from observation. Suppose, instead, $x + x^{1/2} = 1$. Is the solution less than, or more than, 0.5?

Candidate: I am not sure how to proceed. A numerical solution?

Me: Let us not go there. Tell me, do you know the numerical value of $1/\sqrt{2}$?

etc.

3. Me: Please solve $x + x^{1/2} = 3$.

Candidate: $x^1 + x^{1/2} = x^{3/2}$, so $x = 3^{2/3}$.

Me: Thank you. Your interview is over.

(In my line of work, that answer is a deal breaker.)

4. Me: Please solve $x + x^{1/2} = 3$.

Candidate: I am not sure ...

(pause)

Me: Let me give you a hint. It may be possible to turn it into a quadratic equation.

(lightbulb moment)

Candidate: Let $\sqrt{x} = p$. Then $p^2 + p = 3$.

etc.

I believe, as indicated above, that the question should not end with the first answer, but should flow in a somewhat open-ended conversation. Unfortunately, it is still possible to be fooled. Some candidates give you a powerful psychological sense that they are excellent, and this sense is unreliable. Kahneman writes about it at length. It helps if you have committees; and if each committee member gives individual marks on predetermined categories, one by one instead of all at the end (e.g., personality-based questions first, then technical domain questions, etc.).

Sometimes, you will have to choose between one candidate with a track record of very good work and a second, seemingly excellent, but fresh candidate with stunning promise. The first candidate may be the safer choice, but you will find it difficult to make that choice.

You will not feel like going back to check on how your hires are doing compared to how good you thought they were.

The best you can do is be aware of some of the pitfalls, and try to be a good interviewer on the whole.

Exercise 10.

Imagine that a friend you are chatting with tells you he works in industrial safety. He casually says, "Can you suggest some ways to improve industrial safety?" Pause for a moment and fix this setting in your mind. Make it seem real. Are you sitting across a table? Are the lights yellow or white? Are you drinking tea?

Now imagine that you think for a bit and suggest helmets, and he explains he was talking about people in a chemistry lab, not a construction site. He takes a sip of tea and smiles.

You then suggest masks and he shakes his head very slightly. He states that his lab staff already have masks and often do not wear them. He looks at his watch and then turns to smile at you again.

You suggest keeping dangerous chemicals under lock and key, and he says they have that system in place but too many people have keys and the need for frequent access makes people leave containers lying around in the open.

Do you recognize the conversation? Have you had a similar experience where each answer you offered turned out to be unacceptable by some criterion that was not made clear to you in advance?

With this context, look up *bring me a rock*.

As an interviewer, it is not your goal to find ways in which an answer can be rejected. It is your goal to select someone who will be useful to your company.

It is *really* easy to ask questions with unstated criteria of evaluation so that almost any answer can be rejected. Try making up a few bad ones so that you will recognize them when you see them again. I can get you started.

What is India's greatest benefit from having a space program? What is the main advantage of owning an elephant? What is the worst way to sell shoe polish? What is the most critical factor to account for in designing a helmet? Is it okay to wear sneakers to a job interview? Give me a good advertising line for ball bearings. Sell me my laptop. Sell me that flowerpot. Sell me this biscuit. Tell me why I should hire you. Why should I not hire the candidate who interviewed just before you?

CHAPTER 11

Higher Studies

Higher studies within engineering means either a master's degree (usually MTech, sometimes called an ME; and some institutes offer a research-based MS as well); or a PhD, usually the terminal degree.

Many engineering students go abroad for higher studies, and look for jobs there. Many others get master's degrees in India and look for jobs in India. A few get PhDs abroad and then look for jobs, including academic jobs, in India. This chapter is for people with either Indian or foreign higher degrees who wish to build a career within India.

A small number of large Indian R&D organizations, such as ISRO and DRDO, have many PhDs. In many

Indian companies, however, most of the R&D work is done by people with master's degrees and there are few PhDs hired. This is changing, but slowly.

Much of what I have written in earlier chapters continues to apply for MTechs, except that they specialize a bit (e.g., within mechanical engineering, it could be solid mechanics and design, or fluid and thermal sciences, or manufacturing science). People with master's degrees, looking for industrial jobs, are not generalists the way BTechs are. The number of industrial jobs with big companies, at the master's level, is also much smaller than the total number of jobs you could target with a BTech.

I have two main pieces of advice for MTech students who plan to look for industrial positions.

First, it is safer to work for an advisor whose research topic is somewhat meaningful to industry. For example, as a mechanical engineer, you could do something related to stress and fatigue analysis in mechanical components, or you could do something related to, say, nanoparticles injected into a cell in the body so that it responds to radiation in some way. Afterwards, if you apply for a job in a car company, which thesis do you think will help you more?

For example, I had a master's student who simulated a slightly unbalanced rotor mounted using flexible bearings within a box. When the rotor spun, its unbalance caused the box to slide around on the floor (all in simulation: no experiments). He got two jobs: one with a washing machine company, and one with BHEL, which is interested in rotor dynamics. Less satisfactorily, I had another bright student who worked on mathematical aspects of hysteresis, which is relevant to energy dissipation in, e.g., vibrations. However, his thesis emphasized the mathematics of hysteresis and not the vibration damping part (I made the mistake of assuming it was obvious). He had trouble finding a job: many people he spoke to did not think his work was relevant to them.

Second, remember that you have chosen a specialization and are offering a higher technical skill. Employers may be interested in hiring you for design or R&D work. For campus recruitments, some companies start with a written exam at about the level of GATE, say. You probably cleared GATE to begin your MTech in the first place, and you should not have forgotten that material already. Subsequently, in case there is an interview, you must both answer technical questions about your thesis work as well as broader technical questions outside your thesis.

Perhaps you studied a slender solid rotor with a circular cross section, for example. Maybe you developed a three-mode reduced order model and studied a special whirling solution. The interviewer can, in all fairness, ask you about low speed versus high speed rotors, hollow rotors, noncircular cross sections, unbalance, critical speeds, bearings, forward and backward whirling, vibrations, modal analysis of rotating structures, damping, instability, material choice, fatigue, cracks, accelerometers, signal processing, diagnostics, reliability, and so on. If your thesis work included a piezoelectric actuator and a control strategy then – and you must face this – several other windows just opened up as well. Read widely, talk to other students, think hard, watch online video lectures from excellent universities, use commercial computational packages (e.g., for stress analysis or dynamic analysis), participate in experimental studies (even indirectly, in a friend's lab), and generally broaden your technical base.

Understand what the game is. The interviewer does not need to know *all* of these above topics. He or she needs to know a little about just *one* of them. You will ideally know that same one, too. If you do not, you should be able to offer an understanding of something related to it. And if you cannot do that, then you can invite a different question – but you face rapidly diminishing returns. It is a buyer's market. There are few jobs and many applicants. You must work harder, and think more clearly, than your peers. Perhaps, some years later, you will be the interviewer while some younger master's student is in the hot seat. Remember your own experiences, and be kind.

The job market for PhDs differs a great deal from that for BTechs and MTechs. The job search for PhDs is *individual*. There is no campus recruitment worth speaking about.

There are R&D divisions in some Indian companies where some PhDs are hired. Some international companies also have R&D offices in India, and hire some PhDs. I think these numbers will grow, but they are presently small in comparison to the numbers of PhDs graduating from the IISc, IITs and NITs (to mention just a few centrally funded institutions; there are several other nationally known institutions offering PhD degrees).

If you are getting a PhD degree and want an industrial position, then you should perhaps step back from your specialization and try to rediscover general usefulness. The PhD gives you experience with a narrow problem area, it is true. But it teaches you broad and transferable skills, too: problem definition, literature review, planning an attack, working independently, assessing your results, refining your theory or understanding, presenting your conclusions to others and defending your findings. Your task is to communicate this breadth to a potential employer.

For example, perhaps you developed a particular computational model for some microstructural feature of some alloy used in a high temperature application in military aircraft. The interviewer may think that your work has no relevance to the needs of his or her organization. You may need to engage patiently with that skepticism and, given an opportunity, explain that your *skills* are useful to both military and civilian applications, on air and on land and in water, at high and moderate and low temperatures, in alloys and pure metals and composites and polymers, at the level of microstructures as well as larger features like joints, etc., etc. It is your job to widen your intellectual range *and* give the interviewer a reassuring glimpse of the same.

Let me offer a story. There was a PhD student at IIT Kanpur who worked on contact mechanics, which involves stress and deformation analysis in small regions where objects press on each other. For example, if you press two steel balls together, the contact region is much smaller than the balls and the stresses near the contact region are much larger than elsewhere. That student was going for an interview with a company that makes glass. I suggested to him that, before he went, he could do simple bending experiments to break some strips of ordinary window glass in order to estimate failure stresses. He could then use one of our small testing machines to load a similar sheet of glass using a steel ball, shine a laser beam through the contact region, slowly increase the load until cracks appeared (which would be seen because of the laser), and make a video. Finally, he could do an analysis using the bending-breaking experiment and his contact mechanics skills to predict the breaking load in the ball-on-glass experiment. He could compare results and show his video. I thought this would help his industrial job application. I leave it to you to judge whether my suggestion was good[23].

Many PhD students are interested in academic positions. These are limited, too. What follows is advice to such a student, within the specific context of a thesis involving *numerical modeling of multiphase flows* (that is of course a broad area, not a thesis topic). The rest of this chapter is adapted from something I wrote earlier on *Quora*.

To even begin, you must have minimal competence. That means you understand the governing equations of multiphase flow, and the assumptions behind them.

---

[23] As it happens, the student did not implement my suggestion, did not get the job, joined a coaching center as a teacher, and is happy enough. But that is not the point of the story. Suppose you *wanted* the job with the glass company. Would you implement my suggestion to demonstrate an interest in broad usefulness?

When do the assumptions hold, and when do they fail? What do they mean, physically? For example, if you have bubbles in the flow, what have you assumed about the statistics of bubble sizes? Do you have videos to support that assumption? How big must the domain be if you want to make some kind of continuum approximation? Or, are you modeling individual bubbles? How many ways are there of making such continuum approximations? Presumably there is heat transfer. Then what does temperature at a point mean? Is it possible for the temperature in the liquid to be different from the temperature in the bubble? How does the bubble interact with the wall? How do you interpret no-slip at the wall if the bubble gets very close? What is your model for breakup and coalescence of bubbles? If you cannot answer these and similar questions with complete competence and confidence, then you will not get past a good academic job interview. Further, you must understand computation. Issues like two-dimensional versus three-dimensional modeling, serial versus parallel solvers, algorithms that work well and algorithms that do not. You should be clear on which part of the code you wrote, how long the code takes to run, how you would incorporate moving walls, how to incorporate flexible walls, etc. There are many things to think about.

After you are competent in this way, your battle begins.

Imagine you have finished your thesis, written some good research papers, and got an interview call from a good institute.

In the interview they will ask technical questions. They will be sitting, unhurried, with tea and snacks. You may be sitting, or maybe even standing at a board. There is no syllabus. Anything related to your thesis can be asked. Anything related to anything related to your thesis can be asked. They may keep asking until they find questions you cannot answer, because they want to see where your boundaries are. Then comes teaching. All undergraduate material within the umbrella of your subject is in principle included. Basic fluid mechanics, Navier-Stokes equations, Bernoulli's equation (actually tricky, sometimes), fluid statics, flow around a bend, Venturi, ... anything. Compressible flow. Heat transfer. Free and forced convection.

The game is like this. If you make fundamental errors, you look bad. If you make too many fundamental errors, you look terrible: unfit to face undergraduates with only a piece of chalk in your hand. On the other hand, if you can counter an argument using a fundamental principle, or your answers show a good grasp of fundamentals, then you look really good.

One thing PhD students do not seem to do is work on broader fundamentals. They just work on their thesis. I think this hurts them.

Form a study group with fellow PhD students. Meet regularly, something like 6–8 hours a week. Pick a topic, or better, a problem. Discuss how it is to be solved. Let one person present, and let the others think of odd and unexpected questions. The group should agree that the question is feasible in an interview setting, and then the group should evolve an answer until they agree that the answer is good. Occasionally, if you cannot agree on an answer, consult a competent faculty member.

A stand-up, no-syllabus interview.

There is a glut of PhD students and not enough good academic positions in our best institutions[24]. But most PhD students are not prepared for a stand-up, no-syllabus interview. Make that your secret weapon. If and when you reach the interview, you must shine.

Note that some MTech students also take up jobs as lecturers in colleges. They may face a gentler version of the interview for academic positions that I described above, or they may be selected directly based on grades and communication skills, depending on the employer. Such jobs, however, may require getting a part-time PhD in the long run.

Exercise 11.

An exercise. Let us practice moving away from a problem and asking either broader or tangentially relevant questions.

Here is an example from my own subject area. Consider stresses in two steel balls that are pressed together. From there to contact problems in general, stress concentrations in other problems, nonlinear elastic effects, material yielding, what happens when the balls are rolling, ball bearing design, impact problems, role

---

[24] There are teaching jobs available in many lower tier colleges, and you may choose to take one of those jobs. There is certainly a contribution to be made there: many of those colleges need more and better teachers.

of friction, shear stresses in the contact region, contact stresses in clamped joints, stresses in joints experiencing vibration and microslip.

With the above example, choose something closer to your own area of research. Electromagnetics? Power electronics? Digital filtering? Microstructure evolution in special alloys? Battery management? Flutter instability in rotorcraft? Fatigue of composites? Condensation on textured surfaces? Foundations of buildings? Liquefaction of soil?

It does not matter what the starting point is. Think of broader questions, and tangentially relevant questions. It takes practice.

Do you think this exercise will help you prepare for wide-ranging interviews?

CHAPTER 12

Personality and Character

Once you get a job your colleagues will interact with you regularly. They will learn things about you that are not on your CV. Things related to your personality and character.

There is some overlap between personality and character. Here, I will use them for two different things. When I say *personality*, I refer to aspects of your behavior which are seen quickly: things like confidence, a direct gaze, an open smile, a sense of humor, and so on. When I say *character*, I refer to long term aspects of your behavior that are determined by your value system. These aspects are not immediately seen in a job interview or perhaps even within a month on the job. Character refers to ethics, courage, honesty, trustworthiness, humility, patience, persistence, and other similar long-term traits.

Let us begin with personality. Some people have pleasing, confident, outgoing personalities. This can help them in the workplace. In contrast, if you have a shy personality, or lack confidence, or hesitate to speak clearly, then it may hamper your career. Although it is not easy to change your personality, perhaps you can change your behavior in the workplace to some extent. If everybody smiles a lot in your workplace while you do not naturally smile much, you can learn to smile more often. Practice in front of a mirror, remind yourself to smile from time to time, start by smiling at people you meet more often, and so on. If, on the other hand, your colleagues are mostly serious and focused while you like to smile and joke as you work, then you can try to tone down your exuberance.

Let us now turn to character. Since character does not get revealed in quick interactions, you may think that personality is more important for career progression. However, if you engage in a large number of interactions with the same set of people over a long period of time, then your character will be revealed to them. If you want to develop a stable career in a shifting job market, your character may turn out to be important. People who know you well may make allowances for your personality, but they may not do the same for your character.

I know someone who used to work for Coca-Cola, a big company that offers steady career prospects. At some point, he left to start his own business (also in beverages). Several people left with him to join his venture. I believe those people left their safe jobs because they trusted him. Their trust was a testimonial to his character. Such trust cannot be valued easily in rupees. You can read what he has to say, in the form of an external opinion, in the next chapter.

In your own workplace, you might see a few people of poor character doing quite well. Such people may have skills I do not understand, but there may also be connections in the background that you do not know about. In contrast to such people, for typical employees, good character is usually appreciated in most organizations. In some bad environments good character may be tolerated although it is not rewarded. In some rare, highly vicious or corrupt places, good character may be actively punished; let us hope you do not work in such a place.

Regardless of such negative examples, I am in favor of developing positive character traits as a career strategy. I believe it works more often, and for more people. This is because, in the long run, your character is known to your colleagues, both present and past. You can aim to be a contributor who is relied on by superiors, trusted by contemporaries, and consulted and respected by subordinates. Ideally, even if you leave a company, the people who stay behind should continue to think of you as a friend. Life comes around in circles, sometimes, and people can have long memories for bad behavior.

Over time, your character and personality can both be enriched. You can learn much if you are observant and receptive. There are basic skills in the execution of projects and tasks, which you acquire by taking responsibility and doing things well. You slowly understand workplace dynamics[25]. You develop working patterns with colleagues and team members[26]. You learn, from senior people, how to deal with groups and personality conflicts. You learn to monitor and control complex projects. You become better at quickly

---

[25] For example, when a group is given a task, do the group members immediately accept it all, or take one part and shift the rest to others? Do they try to make it clear who is doing exactly what? Do they receive credit for their work?

[26] Maybe one person is punctual while another is erratic but creative. Someone keeps finding reasons to leave while you stay late to finish up. You learn to work with them all.

assessing a new situation before taking key decisions[27]. As you get better at these and similar things, your personality and character evolve in positive ways.

Unfortunately, during your career, you may encounter some people with either unpleasant personalities or poor character. This will lead to conflicts that you will not always win. Sometimes, e.g., with narcissists who are in a position of power over you, even if you win a small battle you may then be punished for it in unfair ways. Narcissists have a poor self-image which they hide carefully, even from themselves; and they have an endless appetite for praise from all people around them. If you hurt their pride in public, they may never forget or forgive. You may have to find another job, or suffer. Similarly, if there is somebody doing something wrong, like misusing power or funds, then if you expose or embarrass them you may make a permanent enemy. You may then have to fight them long term or leave. Fighting an implacable enemy over a long period of time has no glory in it, only fatigue. Even if you win in the end, the price can be heavy.

If you do contemplate such a battle on a matter of principle, then I wish you luck and offer some simple

---

[27] For example, you are moved laterally to head a new group whose supervisor left the company. Or you are asked to step in and save a project that is not making progress.

advice. Assess the circumstances before you proceed. First, what is the relative power balance? If the other person is much stronger than you, or enjoys the support of someone powerful, then conflict may cost you heavily even if you win: be prepared. Second, if the other person is not very powerful but enjoys the support of colleagues because of fantastic personality traits, then you may wish to assess your own position with your colleagues[28]. Third, if the other person brings in more money (e.g., clients or sales) than you do, then your company may face a dilemma – and who knows which way things will play out? For these reasons, before you engage in such battles, it may be wise to know where you stand with your bosses, your colleagues, and your company in general; whether people think your character is straight and strong; and whether you can get another job if you suddenly need one. If and when you do interview for another job, you should not say bad things about your present employer: be gentle and diplomatic if you can.

If you never face such troubles at all, then you should consider yourself lucky. Someone with a broad range of experiences once told me, "People join companies,

---

[28] People are not always rational. Emotions affect decisions. If they like a person, they may resist evidence of that person's wrongdoings.

but they leave bosses." Let us hope you do not face too many such troubles.

If you think these things are unfair or depressing or somehow wrong, I suggest you read the *Mahabharata* and look at the dilemmas it presents. Life is not fully straight and simple, nor black and white. Honesty will not always prevail, the truth will not always come out, and the brave will not always win. Whether you decide to be honest and fair, generous and ethical, reliable and capable, depends on the value system that you build up as you go through life.

To some extent, your character will evolve with the choices you make. For example, if you *choose* to be reliable on a project, then you *will* be reliable. Somebody will notice, give you more work, and rely on you again. You yourself will begin to value reliability. Do not be fooled into thinking this means you will just do extra work while less reliable people will have an easier time[29]. The way I see it, after the work is done and your company has made profits, you will have developed greater ability and confidence. If you ever leave the company, you will

---

[29] It is true, of course, that you will work harder and some others will not. But the story will not end there. You will benefit from having *done* the work. Remember the hypothetical example where everybody knows basic medicine and low-fees doctors are everywhere: if you have a complex medical problem, you will still seek a doctor with an impressive track record of prior work.

leave the profits behind, but you will take your ability and confidence with you.

Remember the wage rate pyramid model and India's demographics. It is better to be more capable and to have a visible track record of having been reliable than to have successfully dodged labor while collecting a salary. The opportunity to *have* a track record of good work done well will not be given to everyone, because there will be too many workers.

In my opinion, for most people in most jobs, good character is more helpful than bad character. Statistically, good character is a good long-term strategy. I believe it is nearly impossible to repeatedly fool increasingly smart people over a long period of time, all the way to the top.

Exercise 13.

An exercise. This one requires soul-searching.

Two children are invited by others to come out and play. The first child is fiercely competitive, a bit rougher than necessary, with a tendency to lie if he can get away with it, and a tremendous focus on winning the game. The second child competes in a fair way, yields a point sometimes instead of arguing hard, does not cheat, wins less often, and smiles even when he loses. On any day in the same game, the first child is more likely to win. But over many months, the second child is called more often to play, and the first child is called less often.

Now there are individual games, and there is also the *set* of games[30]. The second child wins a longer game where the objective is to be called to play as often as possible.

The following famous words are from Grantland Rice (an American sports writer):

> *For when the One Great Scorer comes to mark against your name,*
>
> *He writes – not that you won or lost – but how you played the game.*

Now, finally, for the soul-searching. Which of the above two children would you like to be? Make it a choice, not an accident.

---

[30] I recall a YouTube video where Jordan Peterson called this set of games a "meta-game."

CHAPTER 13

External Opinion 2

Neeraj Biyani used to work for Coca-Cola, which he left to co-found a company called Hector Beverages which, among other things, makes Paper Boat drinks. I got to know him some years ago when they had some mechanical problems in their filling processes and he found my web page at IIT Kanpur. He asked me if I could help. I was willing to try. Luckily, we solved that problem. I have learnt much from him since then, and been useful to him a few more times as well. Over time, he has become a friend.

Neeraj is a highly capable person with great breadth in his understanding of the many things that go into starting and running a business. His professional experience is vastly different from mine. I requested him to give me his views on career planning for this book, and he kindly obliged. He has discussed the role of character in effective leadership. What follows is from him (as written in 2020).

Imagine that you are employed by a well-established company for a salary of 10000 rupees a month. I invite you to join my new company. My company could possibly go out of business within one year, but I offer you 15000 rupees a month. Would you accept?

This question is, more or less, what I asked some people when I started my own company about ten years ago. Those people worked at my old company, and they knew me. About 75% of the people I invited took up my offer. I have spent some time trying to understand why they accepted.

As it happens, my company succeeded. But success was not guaranteed at that time. A reasonable estimate of the chance of success was, I think, 20%. The people who joined me knew this. How did their calculation work? At their steady salary, one year's income was a risk-free 1.2 lakhs. If the new company survived, it would be 1.8 lakhs. If the company lasted for more than 8 months, the income would exceed 1.2 lakhs. But if the company failed, those people might have to look for new jobs.

It is possible to do slightly different calculations with similar results. For example, even if the probability of surviving one year was 20%, the probability of surviving the first two months was close to 100%. The probability of surviving four months was not 100%, but it was high. The average of 100% and 20% is 60%, so with the extra protection for the early months maybe we can think of a higher number, like two thirds. Two thirds of 1.8 lakhs is 1.2 lakhs. Was this what they were calculating?

I asked some of those people why they joined me. They are still my colleagues, as it happens. One thing they told me is that they consider me to be *fair*. What is the economic value of fairness?

The fairness of managers affects their subordinates. In my industry, variable incentives can make up 30% of an employee's income. Note that fair is not the same as *generous*. I would not be paying excessive amounts on average. An unfair manager might selectively grant extra incentives to favored people, at the cost of unfairly low incentives to others. Employees who believe in their own inherent value, and who do not depend on favoritism to earn more money, may therefore prefer a fair manager. They had faith in my fairness, and I wanted employees with high inherent value. The situation suited both of us.

But there is more. There are things that are harder to calculate.

For example, if you earn the same amount of money for the same amount of work, but in one place you feel you have been treated fairly while in another place you feel you have been treated unfairly, will your satisfaction be the same? I think not. Or, if you work harder than your colleague and produce better results, but your manager does not distinguish between the two of you in feedback, will your satisfaction be lower? I think it will.

Look at it another way. Suppose you are a capable employee and have faith in the quality of your work. You are thinking of joining a new company being started by someone you know. That person knows you, too, and has invited you to join the new company. If that person has a reputation for being fair, will it make you feel more satisfied? Or would you feel the same satisfaction even if you thought that person was unfair, or maybe even incompetent?

In any case, I was fortunate that so many of the people I invited agreed to join me. My company would not exist without them.

Over many years, I have concluded that character helps business succeed. In particular, there are four attributes that I have found valuable in building rapport with colleagues, managers and team members. I list them using the acronym HERO.

1. HONESTY.

The fairness I discussed above is a part of overall honesty. I believe in letting my employees know, gently but clearly, where they stand with me. If they are doing well, I let them know. If they are doing poorly, I let them know too. If it seems things cannot work out, I give them some time and offer to help them in looking for other options.

2. EMOTIONAL RESPONSIVENESS.

Employees have lives outside the company. They have family issues, health issues, and other ups and downs. My knowing something about them helps all of us. Employees come to the company to work. But if we pretend that the rest of their lives do not exist, and we reduce them to machines, then we have poorer outcomes in the long run.

3. REMAINING HUMBLE.

Remaining humble is a conscious process. As our success grows, and as we have more employees in the organization who follow our instructions, we must pay attention to remaining humble. My success rests on my employees, naturally. But it is more than just their work. They often know things that I do not, and I want them to trust me and tell me. If I am arrogant, they will not offer their opinions.

4. OPEN-MINDEDNESS.

By being open-minded, I mean being receptive to unexpected ideas from unexpected sources. Individual operators of my equipment know things that my managers do not know. People outside the company can say things that impact the company's working. Future opportunities may emerge from casual conversations, if one is receptive. I think being open-minded is somewhat related to remaining humble. Each may depend on the other.

What I have written above is a summary of my philosophy of effective leadership. However, I believe it has something to offer to both the leader and the team member. It is a lens through which a worker may like to assess his manager. HERO leads to higher trust and better outcomes for both manager and workers.

CHAPTER 14

Failure and Persistence

Much of this book has focused on building value, getting jobs, and doing well at them. But India's demographics cannot be denied. Many young people will not get jobs. Most engineering students who do get jobs may not get decent jobs in engineering, and they may have to search long and hard before they find something. Many others will eventually shape their own careers without getting steady jobs at all. If you are one of those people, ones who are not getting a job and do not know what to do, then this chapter is for you.

The first thing to do is to prepare *emotionally* for failure. Failure is many things to many people. In particular, many upbeat writers point out that failure is an opportunity: to try again, to succeed next time, and so on. In fact, a Google search for

failure opportunity

returns hundreds of millions of pages. But failure is also a burden, and it can crush you if you are not prepared.

This reminds me of my college friend Jayant. In those days, around 1987, even simple roller skates were interesting things for us, and someone had brought a pair to our hostel. Most of the boys in the hostel did not know how to balance on those skates. The skates were shared widely. Day or night, someone was wearing them somewhere in the hostel, holding on to a pillar or furniture, his face tense with concentration. And then, to our surprise, Jayant was moving about expertly within three days. We asked him how he did it.

As it happens, Jayant was physically very fit. He told us, "I realized that fear of falling was stopping me from learning. So, I practiced falling first. I fell like this. Then I fell like this ..." As he spoke, he demonstrated: falling forwards, backwards, sideways. "After I stopped being afraid of falling, I learned quickly."

For thirty years, that moment was just an amusing memory for me. Jayant saying, while repeatedly falling, "*Pehle hum aise gire, phir hum aise gire, ... Jab girne ka dar chala gaya, tab hum seekh gaye.*"

Recently, on the internet and elsewhere, there are many people talking about a *hundred rejections*. The idea is that rejection feels bad, but we can make ourselves less sensitive to it. We can go out every day with the *aim* of asking people for things that they are not likely to give us; and we can keep count of the number of times we get rejected.

The idea is remarkable. First, and most simply, we can practice getting rejected at things we do not care about, like asking a random stranger for money, or asking some random house owner if they will let us plant a flowering bush in their garden, or asking some random flat owner if they can take care of our potted plant for a week, or asking a random taxi driver if he will let us paint a blue circle on his cab door, and so on. This is different from asking a stranger for a much-needed job, hoping that we will get one. Here we are not job-hunting. We are not desperate, and do not really expect the stranger to give us what we want. The emotional stakes are low. Some challenges remain, of course. We still must go through the difficult tasks of walking up to a random person, making eye contact, finding an instant where something can be said, and asking for whatever it is.

Second, we can try to extend the conversation. Maybe we can ask, "Why are you unable to help me?"

Or, maybe we can ask, "Do you know anyone who can give me what I want?"

Many people have found that getting rejected a hundred times has changed them in powerful and useful ways. Maybe it will work for you, too. If you are unable to get a job, and are experiencing rejections, then maybe you can reduce the emotional burden of it by first practicing getting rejected on small things. This is like my friend Jayant falling deliberately, for practice, before learning how to roller skate.

But you can do something a little different, and maybe even a little better. Let me first explain my reasoning, and then consider specific actions.

When we talk of many failed attempts leading eventually to one success, we can think in terms of probability. The difficulty lies not in probability theory as a subject, although that is difficult enough. The difficulty lies in knowing what the probability is. Let us consider four problems, arranged in order of increasing difficulty. In one we know the probability accurately; in the next, we do not know the probability very well and must guess; in the third, we cannot guess the probability accurately but can still guess the rough magnitude of a trend; in the final one, we are not even able to guess the magnitude of a trend, though we know which way it goes.

Problem 1. Suppose you toss a coin. What is the probability of getting heads? Even with a slightly imperfect coin, it is close to 0.5. You know that. But even if you did not know it, you could toss several coins several times each and *find out* that the probability is close to 0.5.

Problem 2. Suppose I tell you that I recently saw a grown elephant in some other city. It was standing next to a car. I then ask you for the probability of *that* elephant weighing more than *that* car. You have not seen either the car or the elephant. Some big cars weigh more than some elephants. But typical grown Indian elephants weigh more than typical Indian cars. You must guess a probability without complete information. Maybe you guess the probability to be "more than 0.8" while someone with more information says "more than 0.9." It does not matter: the difference between 0.8 and 0.9 is not that big for some purposes. My point here is that experiments with elephants and cars are not as easy to do as with coin tossing. You could go out and weigh some elephants, or you could use the internet to find out the weights of some cars. You would find there are *some* cars that weigh more than typical elephants. But you still would not know which car I saw, and in which city; and whether the elephant was in good health. You would not know whether I am asking the question *because* it was a huge car and a thin elephant (i.e., the answer may depend on what sort of person I am). This problem, with the given information, simply cannot be resolved as precisely as the coin toss problem.

Problem 3. We discussed the UPSC exams earlier, where the population success rate is one in 1000. Suppose I tell you some girl from some village sent me an email and said she is appearing for the UPSC exam. What do you think her probability of success is? You do not know which girl, and which village. You are not even aware of the relative success rates of boys versus girls, but you suspect that it may be slightly poorer for girls. You know that candidates from cities may have better chances than candidates from villages. So, you suggest one in 2000. At this point, I start giving you more information. I tell you this girl has studied in a college in a nearby town, and scored in the top one percent of students in her university. Now you may revise your estimate to one in 100. I then tell you, further, that she has joined a famous coaching class, has been studying there for 2 years, and has been doing well. Now you may revise your estimate to 1 in 20. Clearly, your estimates are not very reliable. Yet, one in 20 is a lot better than one in 2000, and that *trend* in your estimates is fairly reliable. With increasing information, you have been able to revise your probability estimates in a reliable *direction*. A key point here is that the girl remains the same and her true probability of success remains the same, independent of the amount of information we work with. Yet, we cannot really know her probability of success. First, there is the huge range. There must be some UPSC candidates who are so disadvantaged or poorly prepared that their probability of success is one in 20,000. There may be some who are so extremely well prepared that their probability of success is one in 2. That range is a factor of 10,000. Here, if our estimates place us within a factor of 4, we have done well. If the true probability is one in 50, and we have guessed something between one in 12 and one in 200, we have done all right. Think again of the elephant. The weights of grown elephants may vary by a factor of 2, say. They cannot vary by a factor of 10,000. Moreover, we can in principle weigh 100 elephants, but we cannot ask the girl to appear for the exam 100 times; and if we study the performance of 100 different girls we do not know how similar they were to begin with. In this way, the third problem is much harder than the second one[31].

---

[31] One can in principle do better with huge amounts of data. Imagine a coaching institute that has been training 50,000 students a year for the last 20 years. For each student from each year, they have internal records of performance on 50 internal exams, and they also know whether the candidate finally succeeded. They could in principle develop detailed statistical models which predict each candidate's probability of success after the coaching period. Of course, they may not reveal their findings, or they may report overoptimistic estimates to the candidates for various reasons.

Yet, with more information we could still improve our estimates.

Problem 4. The final problem is the hardest of all. I introduce you to a typical student of some typical small engineering college. He graduated three years ago with below average marks, studied for GATE but did not do well, and has been looking for a job for one year. His English is not very good, he lacks confidence, and his interview skills are below average. What is the probability that he will find a reasonable job within the next year? You may say it is small. But how small? There are about 25 million Indians of his age, most of them without good jobs. Surely his probability is better than one in 25 million. How about the almost ten lakh people who appeared for GATE with him? Are his chances one in ten lakhs (one in a million)? Maybe it is higher. Should we look at the students who studied in his college with him, in his year? Maybe 5 percent of them got jobs. Are his chances 5 percent? Probably lower than that as things stand, because he is one of those who did not get a job then, and spent time doing other things that did not work out. I think we cannot easily make even an initial guess about his probability of finding a good job. However, we can still say some things. If he spends most of his time moping at home, or going round the same circle of acquaintances who have been unable to help him in the past, then his chances remain low. On the other hand, if he keeps trying new ideas to increase the breadth of his search, his chances improve. Maybe he extends the circle of his conversations to acquaintances of acquaintances; goes to other towns to visit relatives and engages in career related conversations with a new set of people there; and develops a better understanding of the job market so that he has a more realistic view of where his chances are better. Then his chances improve. If he persists in this way for some time, then his chances improve further. We do not know the magnitude of the trend in increasing probabilities, but we know which way the trend goes.

Let us return, now, to you: a candidate who has failed so far to find a job, is psychologically prepared to bear failure for a while longer, and is ready to strengthen himself or herself by facing a hundred rejections. What are some things that you might do?

Here are my suggestions.

You can ask random strangers for advice on work in general. Do not ask for a job, and do not ask people who have jobs to give. That lowers the pressure for both people. As you approach the person, you can open the conversation in different ways. For example, you can say that you are a student and you are conducting a survey to find out what sort of work people in society need done. Electrical repairs, grass cutting, mathematics tuition for schoolchildren, polishing of furniture, it does not matter what they say – take notes and slowly build a picture in your mind. As another example, you can walk around, find someone you know who is doing something, offer to help, and *after* the work is done, ask for advice on job hunting. You can ask that person for the contact information of somebody else who may also be willing to speak with you. And so on.

Your effort counts if you manage to have at least a few minutes of conversation on any general topic related even loosely to work. If you can have a hundred conversations like this, it will change you and make you stronger. And there is a good chance that it will also give you some ideas about finding work (maybe single tasks or small contracts), about skills you want to develop, about starting new businesses, or even just an introduction to someone who might point you in useful directions.

Imagine that you have already had 50 such conversations. Then, quite unexpectedly, one of these people tells you there is a vacancy in some company where they know someone. You send in your CV and you get an interview. Think about how much more confidence you will have during that interview. Such confidence cannot be built by taking vitamins, doing pushups, or reading books.

Or, imagine that someone tells you about some startup nearby which is making a type of product which you have discussed with other people. You would feel less shy about going to look, and to talk. Who knows what that may lead to?

This does not necessarily mean that you will stop failing. This only means that you will search out more opportunities through random connections, and be both more realistic and less shy in pursuing those opportunities when they present themselves.

Beyond this, you must remain persistent. That is why I suggested one hundred failures, counting only the ones that result in at least a few minutes' conversation. That process can take a few months.

Let me offer some mathematics. Imagine that any such conversation you engage in has a small probability, like 1%, of leading to something that leads to a career. Then the probability of leading to nothing is 99%. But if you have a hundred such conversations, your probability of failure is only about 37%.

Reality is better than that. If in fact you have a hundred meaningful conversations, then the probability that the next conversation will lead to something useful keeps going up. This is because you become better at talking to people, those people point you selectively to other people who might be willing and able to help you, and you become better prepared to detect an opportunity when you glimpse one.

Keep your faith. It is a big country. There is a lot of work that needs to be done, hiding within the population, and there are not enough reliable and capable people who are willing to *find* the work and *do* it.

Exercise 14.

An exercise.

Read online about *six degrees of separation*. This is a remarkable idea that says that you could take almost any other person in the world, and the two of you are probably separated by a chain of not more than 6 acquaintances. Of course, this is not a mathematical theorem. The actual number may in some cases be larger than 6, but it is often surprisingly small.

Can it be true? For a difficult-seeming example, how long is the chain from some random village boy in India to Lionel Messi?

Think about possibilities. A postman probably comes to the village and the boy may know him (that is "one"). That postman may have met a more senior postal officer on some trip to a big city (two). That senior postal officer may once have been ill, and met a senior city doctor who has many famous patients (three). One of those famous patients may know somebody very senior in the Indian government (four). That senior person may have met an Indian diplomat posted in Argentina (five). The Indian diplomat in Argentina may have met Messi directly (six), or may at least have met someone who has met Messi (seven). This may not be the shortest path. The boy may go to a school where there is a teacher who studied in a big college in the nearby city (one). That teacher, in his student days, may have met some senior professor who was visiting (two). That professor may have travelled to an international conference and met a European professor (three). The European professor may have spent some time in Argentina and met somebody well known there (four). And so on.

I repeat: it is not a mathematical theorem. But the idea is remarkable.

After this exercise, as a job seeker, you may agree that there can easily be a chain of acquaintances from you to hundreds of people, maybe 5 or 6 handshakes away, who can help you with your career. These days everyone looks online, and you have probably tried that already. Now try a chain of people. Six may be all you need. But *which* six? This is why you must talk to many people; and broaden your search a lot; and not pressurize people along the way. Ask for advice and information, not a job. Somewhere along the way, a job may be waiting.

It may help if you are capable and reliable. Then at each link on the chain, the person you are talking to may be more inclined to send you on to another, possibly helpful, person.

Repeat after me: *Somewhere, six handshakes away, is a person who can give me a job.*

CHAPTER 15

Making Your Own Luck

Fortune favors the prepared mind[32]. This means that people who are more ready can detect better opportunities more often. They seem to have better luck. This chapter is about improving your luck.

Many people misinterpret luck. For example, consider the JEE Main, where you get a percentile score. Someone might take it twice, getting a score in the 97th percentile the first time and in the 92nd percentile the second time. Usually, they will say, "I had bad luck the second time."

But they could be wrong. Maybe 92 is their correct score, and they had *good* luck the first time. Maybe their

---

[32] From Louis Pasteur, a great scientist who saved very many lives.

true score is 65, and they had amazingly good luck both times. People tend to attribute their successes to their own qualities (like talent and effort), and tend to blame their failures on luck. There is social pressure in this direction, too. If somebody fails at something, we politely say, "Bad luck." In some cases, "Well deserved!" might be more honest, but it is socially unacceptable.

Not all luck is the same. Perhaps you know someone who won a big lottery. Such luck cannot be copied. Most people who buy lottery tickets do not win. And if winners buy again, they tend to lose their money just as much as anybody else. Buying a lot of lottery tickets is, statistically, a losing strategy. I am not interested in such luck as a strategy.

Here is a different situation. A young man I know was visiting a friend's house for a meal. Another visitor there was offering that friend a business opportunity: zonal distribution of pharmaceuticals made by some small company that was trying to enter that region. The friend hesitated, but my young acquaintance asked for details and took up the offer. As it happens, the business did well. Hearing that conversation from the side was a bit of luck for this young man. Such luck is not the same as winning a lottery.

Note that the initial offer was made to the friend, and not to my acquaintance. The friend could have taken up this offer, too. Since the business succeeded, the proposal was probably not a bad one. Then, why did the friend hesitate while my acquaintance took it up? I think the first person was not prepared, and the second one was. He asked how much money would be needed to get in, and he knew that he could find that much money; he had thought about this and other business opportunities enough to know whether he wanted to take a chance on this one. In this way, luck favored him because he was prepared.

Of course, the business could have failed, and then I would not be telling you this story. When you hear success stories, watch for *survivor bias.*

I was once talking to a PhD student of management. He said he was trying to identify traits that make companies successful. His plan was to interview senior people from some highly successful companies, and ask them what traits their companies had which led to that success. The problem with this approach is survivor bias. Imagine that, after interviewing these people, he discovered that these successful companies (i) took risks, (ii) had a flat management (i.e., not many layers of middle management), and (iii) trusted their employees rather than monitoring them closely. These traits seem excellent. But are they?

To understand the difficulty, suppose that you want to find out if snake bites can kill people. You go to a big railway station (say in Mumbai), and interview one million people one by one. You ask them two questions: (i) were you ever bitten by a snake, and (ii) did you die?

You might conclude, with false statistical rigor, that snake bites do not kill. Your sample was not representative of the population that you are truly interested in. You did not interview the dead people: hence, survivor bias.

Back to the PhD student. He did not plan to interview companies that did poorly. Maybe *many* companies took risks, had flat managements, trusted their employees, and went bankrupt. They would correspond to the dead people in the snake bite survey. There are other problems. Maybe managements of these successful companies merely *think* that these three traits led to success, while the real reasons were different[33]. Maybe there were not even clear reasons, and luck played a big role[34].

---

[33] Timing, tax breaks, political links, … many possibilities. Also, think of the opposite experiment: interviewing the managements of companies that did poorly. Will they say that they failed because they (i) did *not* take risks, (ii) did *not* have a flat management, and (iii) did *not* trust their employees? Or will they offer other reasons?

[34] Professor Joseph Cusumano of Penn State once asked me this. Imagine that 100 soldiers were sent to destroy an enemy machine gun on a hill. 99 soldiers were shot and killed, but the machine gun was captured by the last remaining soldier. The question is, was he the best among those 100 soldiers?

With the above introduction, we can now think about why some people have more luck than others. Sometimes it is like winning a lottery, which cannot be copied or repeated. Sometimes it is like my young acquaintance, at the right place and time, taking up an offered business opportunity because he was prepared. And sometimes it is like the PhD student looking at successful companies, where our prior frame of mind constructs causal explanations and keeps us from even noticing the role of luck. Of these three, the second case is an example of fortune favoring the prepared mind.

What is a prepared mind?

For the present purposes, think of your mind as an extremely large network or structure made of things that you know, along with connections between them. The things you know include people, places, ideas, facts, theories, methods of analysis, patterns in data, shortcuts for specific problems, and so on, and on – a gigantic set of very different items. These things do not exist in isolation, like a large collection of distinct dots. They are connected in deep ways within our brain. To offer just a tiny example, the equation

$$x^2 + y^2 = 1$$

may be connected in your brain with circles, but circles are connected to donuts, the Olympics, earrings,

Mahatma Gandhi's eyeglasses, a pie chart you saw in an article, a cup of tea, and so on. And the donuts are connected to other things, and Olympics are connected to yet other things, etc. Meanwhile the equation itself may be connected to quadratic equations, derivatives, tangents, triangles, trigonometry, the imaginary number i, and other such mathematical things. This staggeringly large network or structure of interconnected things is a useful model for your mind.

When you are young, you learn quickly. But you also know fewer things, and the structure that makes up your mind is smaller. As you grow older you cannot learn as fast, but you know more things and your structure is bigger.

Psychologists talk of two kinds of intelligence: fluid intelligence, which is the ease and speed with which you can process new things in creative ways; and crystallized intelligence, which is the ease and speed with which you can use what you already know to do useful things in a given situation.

Your fluid intelligence is whatever it is. I cannot help you increase it. But it starts going down past age 20 anyway. Your crystallized intelligence lasts well past middle age. If you are 20 years old today, then the age of 50 may seem very far away. But that age will come one day, and you may live for another 30+ years after that.

Over such longer time frames, it seems useful to think of your mind as a big structure with many connections. That structure is the playground of your crystallized intelligence. You cannot build it in a day. You build it over years of being curious, talking to people outside your work, thinking outside your specified tasks to discover connections, watching clever people do original things and trying to sense their approach, taking questions from exams and changing them and changing them again, helping other students with their studies and getting glimpses of their thinking, and so on, and on. Do not just accumulate dots (as in facts). Work on connecting them.

Our present educational system does not, in my opinion, do a very good job of building such connections. A traditional engineering education does slightly better than some other subjects. But the system cannot be blamed. Teaching in a typical school or college in India is a hard job, with long hours, much grading, detailed syllabi, crowded classrooms, and moderate to low salaries. Give your poor teachers a break, and take charge of your own higher development.

Let me offer an example of a connection. When I was teaching at IIT Kharagpur, a group of us visited RDSO in Lucknow. On a lab visit we saw a wheel braking system test setup, with thermal cameras.

A different problem, mentioned in a large meeting by another senior person, was wheel gauge widening, wherein the wheels of the locomotive spread slowly and slightly apart (with a risk of derailments). As a student of mechanics I knew something about creep, which is a slow viscous-like response of loaded metals at high temperatures. The connection that occurred to me was that under heavy braking the wheels may heat up enough for creep to occur; and if creep occurs, then the wheel gauge can widen. My young friend and highly competent colleague Vikranth Racherla took up a project with the railways, did interesting experiments (both in labs and on trains), and found many more complexities in the problem. From my perspective, it was a matter of luck that those two conversations occurred in the same organization within several hours; it was further luck that I knew a bit about creep; and it was a final piece of luck that, in that place on that day, I realized there was a connection. Maybe there were other connections that I did not see, including some that were seen by others. But the one that I *did* see – I offer that as an example of fortune favoring the prepared mind. You may say that the insight was obvious, and does not deserve a paragraph in this book. You may be right, but there were several people in that room on that day.

How can you build up your own network in your mind? This is hard for many reasons. First, there is no unique list of things to learn. Second, the task is very big. Third, it is not clear when it will pay off, or in what way. Fourth, if you do not build it up and thereby miss an opportunity, you will never know what you missed. All you may do is observe others who make connections and grab opportunities, and think that they are lucky.

If you do not find it rewarding to keep working on building up such a network, you will probably not do it. You may need a reward mechanism. The reward need not be big, it need not be steady and deterministic, and it can be social rather than economic or professional. If you are a naturally curious person, maybe you will enjoy thinking about varied topics and seeking connections anyway. If you are a sociable person, maybe you can find a small group of friends and turn it into a game. The game could involve unstructured searching, some analysis, interpretation, and a presentation of it all to the group.

It could be simple, like "What is the power consumption of an escalator in a mall?"

It could be more open ended, like "How close to the edge of a cliff can a heavy bulldozer go?"

It could be offbeat yet difficult, like stress analysis of a sofa cushion. It could be practical, like cost estimation of mass-produced buttons or belt buckles.

If you are prepared to actually put in effort, consult people, use software, read books, watch related video lectures and demos on the internet, talk to designers, and then critique each other as you present your findings within the group, then you will be transformed. Most people will not put in this kind of effort. Those who do will become better than those who do not.

There is another side to the process: making it more likely that opportunities will continue to appear. This aspect is discussed widely by many others, so I will state the ideas briefly. You need to get out of your comfort zone, meet people who do work different from yours, offer to help people and learn from them as you help them, and develop a large set of contacts who trust both you and your ability[35]. When you do a piece of work, do not look at the money unless you really need it. Look, rather, at what it will teach you and whether it will help you find other work later.

I have a good friend and former classmate from college called Shyam. We studied mechanical engineering together in IIT Kharagpur. Then he went to the US. He got a PhD from a prestigious university, worked for a big car company, got an MBA from another prestigious university, worked for a big

---

[35] I discussed this at some length in the chapter on failure and persistence.

consulting company, and eventually started his own small consulting company. He is doing well. He now has offices in Chicago and Bengaluru. Shyam tells me the following:

You get work from people you know. You work hard, and you are honest with them. If you are sure of your result, you give it to them. If you are not sure, you tell them so and apologize. You never give them something with false guarantees, because if it fails and they know you lied, they will not tell you anything. They will simply never give you work again.

Imagine someone saying, "Shyam is lucky. He found some customers who keep coming back to him. Initially, I found some customers too. But they never came back. I have had bad luck."

You do not want to be that person. Build a reputation for both ability and character. Luck will follow.

Exercise 15.

Time for an exercise. Watch a couple of online lectures and talks by Nassim Nicholas Taleb on being *antifragile*. The introduction below will help you prepare.

Antifragile is not the same as robust. Robustness is merely the absence of fragility. Antifragile is the active opposite of fragile. Taleb has written a book on the topic, but the book is mathematically sophisticated and I will avoid details.

He says important things using ideas of fat-tailed distributions in statistics. My informal summary of it is this:

> *Life is long, multidimensional, and complex. The future contains events that you cannot anticipate, far less predict. No amount of watching the past will let you predict them. You must prepare for them in a different way: qualitatively rather than quantitatively.*

Taleb's non-specialist example of fragile behavior uses a coffee cup. If you manage to drink your coffee and put the cup down, your benefit is "1 coffee." The random variable here is the care with which you put the cup back down. If you are distracted or in a hurry, you may become careless and break the cup. The chance of breaking the cup is small on any given day. This is an example of small positive outcomes in many trials, but a lurking small probability of a big negative outcome which could come without warning. Taleb calls this system *fragile*.

The idea is powerful, but not scary because the potential loss is not severe. Breakage is a big thing for the cup, not for you. You have other cups.

So, think of the game called Russian Roulette. In this game you load a single round in a revolver, spin the cylinder, point the gun at your head, and press the trigger. If you survive, somebody pays you money. Now unless you are drunk, suicidal, or under compulsion, you are unlikely to play the game. But the probabilistic parallel with the coffee cup is clear. This set of circumstances makes you fragile. Even if the revolver has 100 chambers instead of the classic 6 chambers, you may hesitate to play.

Compare Russian Roulette with going out without an umbrella during the monsoon season. You do not know if it will rain, and even if it rains you do not know how much it will rain (assume you do not watch weather news). You can choose the assured convenience of not carrying an umbrella, with the potential inconvenience of getting wet if it does rain. If it rains a lot, you may get very wet, but you will not drown. Here you are not fragile. You are *robust*. There are no big rewards and no big penalties, just a gently rising cost against the intensity of the future circumstances.

There is a third type of situation. Taleb suggests that you think about, and indeed seek out, situations where there possibly is a small cost to you on most occasions, but there is a small but reasonable chance of a big positive payoff. This situation is more than robust. Taleb calls it *antifragile*. His examples are often geared to stock markets, but I offer an interpretation for normal living.

When we invest 100 rupees and get back 105 rupees, we call it a 5 percent return.

It may seem stupid to make an investment whose return is *minus 95 percent*. You invest 100 rupees, and all you get back is 5 rupees. But not all rupees are the same. It matters when the rupees go out, and when they come in. It may be that you pay one rupee a week for 20 years without feeling any burden, but on one cold wet night when you have lost your wallet, you find a warm dry place where the entrance fee is 50 rupees. Circumstances matter.

What does this mean for life? Every time that you can help someone at only a small cost to yourself, *please* help them. Maybe 95 percent of the time it will lead to nothing. But life is long. A day may come when you are caught in a peculiar situation, and one of your old acquaintances will step forward and do something that you direly need. It has happened to me[36]. It could happen to you. What is more, even if it does not work out, you will have lived a better life, not felt much loss, smiled more often, and left the world a better place.

Now, go to the original source and listen to Taleb on being *antifragile*.

---

[36] An example is a former PhD student of IIT Kanpur. He was from another department, and notionally I had nothing to do with him. One day he wandered into my office seeking help with a couple of equations. IIT Kanpur has many competent people, but he had wandered a bit before he found me. I suppose others were busy. I spent half an hour on his equations, and I helped him as I have helped many other students. I thought no more of it then. Several years later, he still remembers. He has meanwhile risen to a senior position, and more than once made a difficult task easy for me by talking to somebody useful. Sometimes even proper things get stuck, and a little help goes a long way. I was lucky to know him! And here is my point: to some extent that luck is a consequence of a minus 95 percent investment.

CHAPTER 16

Looking Forward

We learn from the past, live in the present, and plan for the future. Since the future is unknown, we plan without full information. Much has been said by many people about such planning. Here is my understanding of the matter.

To understand randomness in the future, an understanding of probability can help but it is not enough. Some writers distinguish between risk, wherein probabilities are known, and uncertainty, where even probabilities are unknown. Some people talk about *known* unknowns versus *unknown* unknowns, which roughly refers to things you can plan for and things you cannot.

An example of a known unknown is the exact amount of rainfall we will receive next year. An example of an unknown unknown is the recent Covid-19 pandemic – in the summer of 2019, most people did not imagine such a thing coming so soon.

Do not be fooled by hindsight and think *you* knew Covid-19 was coming[37]. Maybe you dimly perceived the scope of a large-scale disruption. But there were many other candidates. I suggest, e.g., large asteroid impact, a huge locust swarm that eats crops in several countries, intercontinental drift of a fungus that kills crops where it goes, international spread of something a little worse than Ebola, several near-simultaneous large terrorist attacks on the world's financial centers, several tsunamis within one year, and a global computer virus attack that affects millions of offices within a week. Which of these do you think will happen first? What is the probability that one of these will happen within the next two years? What is the probability that the next big disruption will not be one of these but something else entirely?

Yet, even though we cannot know the future, we must plan for it. For such planning, beyond trying not

---

[37] I recommend reading *The Black Swan* by Nassim Nicholas Taleb.

to worry about the very big things you cannot control at all, I offer three suggestions.

The first suggestion is to remember is that life is high dimensional. The set of all future possibilities is too large to be fully understood, but it is likely to contain positive outcomes if we are ready for them. Let me offer an example.

Several years ago, I had a research grant and wanted to hire a project assistant. The salary was small. I advertised in a newspaper. Among the applicants was a young man named Vishal.

Vishal was a mechanical engineering graduate from a nationally renowned institute, about 7 years out of college. After graduating with good grades, he had worked for a big automotive company. Then he appeared for the CAT, left his job, and joined the MBA program of a well-known institute. There were three streams in the MBA: finance, consulting, and production. Because of his background he took the production stream. When he graduated, there was a slump in the manufacturing sector. He did not get a job in campus. Unfortunately, at that time, his father fell ill. He left his institute and came home to take care of his father. After some months, his father passed away. So, he stayed on at home, looking for a job while taking care of his mother. He did not find a job for several months. His savings dwindled. Finally, he saw my ad and applied for a temporary position where the salary was one fifth of his old salary.

Now, as far as I can see, Vishal did not make any obviously bad choices. He got into a renowned undergraduate college, got good grades, got into a renowned company, got an MBA from a very good institute, *and* he took care of his parents. Yet, things turned out poorly. His perfectly sensible plans had not worked out.

But life is high-dimensional. Things did improve for Vishal. Within a few weeks after I hired him, he got a job with another company in Kanpur at about two thirds of his old salary. He asked me if I would release him (that was decent of him). I encouraged him to take the new job. As it happens, he left that company after some months and took up mathematics and physics coaching because his brother is in that same business. Some years later, he joined his brother's business. He tells me that he is happy, and doing well.

Notice the *other* possibilities. He might have stayed on with that second company, and moved later to yet another company. Or, if he did not get that job with the second company, he might have stayed on with me and done something else, like enroll for a PhD in our management department (which is possible after his MBA). Or he could have done some good technical work on my project, and applied abroad for higher studies. Something would probably have worked out.

A good way to see that life offers big and unexpected changes over decades is to look at older people. Since I am such an older person, look at me if you like. Today, I teach engineering at IIT Kanpur, and expect to retire from here. But, a little over ten years ago, I was teaching at IIT Kharagpur, and expected to retire from *there*. Ten years before that, I was teaching at IISc Bangalore, with no plans of leaving. Ten years before that, I had just landed in the USA as a student, with no clear plans of returning. Ten years before that, my father had just passed away after a serious illness, and I had every intention of studying medicine. And ten years before that, I was very young and had no long-term plans.

So, in looking forward, my first suggestion is to remember that life is high dimensional. The future will probably provide unexpected opportunities. As we face the future, a bit of faith is a good thing.

My second suggestion is in the area of how we make choices.

Here are some simple questions. As you read them, think about why they are simple.

Is it better to study computer science in a small college without a big reputation, or chemistry in an old and well-known college? Is it better to stay in your automotive job of the last four years, or to enroll for an MBA in a well-known institute? Should you join the Hyderabad design center of an American company, or should you take an Indian factory job where you will earn less but the power structure is more local? If you want to take an online course in the evenings, should you take machine learning or German? Such questions do not have unique answers. People can find reasons both for and against each possible choice. Another person's right choice may not be *your* right choice.

I think the above questions are simple because each one offers just two options. Of course, the above binary choices are not as simple as the binary choice of whether to have pizza or *biryani* for dinner on some given night. The pizza versus *biryani* choice is fully understood, has small immediate consequences, and no long-term consequences. You could toss a coin to decide, and your life would be no different either way. The binary choices in the previous paragraph all involve an unknown future with significant long-term consequences. Nevertheless, they are binary. Binary is simple.

The problem is that many choices in life are far from binary. Sometimes it may be difficult to even state what the choice is. Such situations are complex.

For example, suppose you are from a small town and just graduated from college with a single job offer. The job is not very good: it involves long and hard hours, poor future prospects, and low pay. Should you take up the job? And if you do *not* take it up, then what should you do? Should you stay at home and keep looking online? Go and stay in a cheap hostel in a big city, in the hope of finding more local opportunities? Take a six-month course in machine learning, or a twelve-month course in data science? Take GATE coaching for the remaining months until next year's exam date? Start preparing for the UPSC, which can take three to five years and offers a low chance of success at the end? Start your own business? Join an NGO that is doing interesting work in efficient farming techniques? Seek a project assistant position in a research lab? Enroll for a self-financed masters in a private college? Consider some other exams, like the GRE, the GMAT, or the CAT? Start tutoring local schoolchildren or JEE aspirants or some other such group of academic aspirants?

I do not think it is possible to find a meaningful mathematically optimal solution in such situations. There are too many factors to consider. Many of the factors are not precisely known, and their roles are not precisely quantified. In such situations it is better to use simple heuristics to clear the clutter. To leave fewer choices on the table. To simplify.

To see how simplification helps, consider the above choices as an initial list.

1. Take the job.
2. Stay at home and keep looking online.
3. Go and stay in a cheap hostel in a big city, to look locally.
4. Take a six-month course in machine learning.
5. Take a twelve-month course in data science.
6. Take GATE coaching until next year's exam date.
7. Start preparing for the UPSC.
8. Start your own business.
9. Join an NGO working on farming techniques.
10. Seek a project assistant position in a research lab.
11. Get a self-financed masters in a private college.
12. Consider other exams, like the GRE, the GMAT, or the CAT.
13. Start tutoring local schoolchildren or JEE aspirants.

The list is too long and varied. The future possibilities arising from each choice are not rationally comparable. There is too much choice and too little knowledge.

Let us simplify.

Suppose your family circumstances are poor. You need some money. You need it soon. This suggests that you cannot take long-term non-earning paths that do not guarantee eventual financial success of some sort. Choices 7, 8, 9, 10 and 11 then seem poor, bringing your list to eight items. Still too many. Choice 6 may not result in a PSU job, but might lead to an MTech, after which you still must fight it out in the job market. Meanwhile, you need money. Would you like to drop that? Choice 12 could lead to higher studies abroad … but will your budget and time frame allow it? Suppose you drop that, too. Then you have the shorter list:

1. Take the job.
2. Stay at home and keep looking online.
3. Go and stay in a cheap hostel in a metro city, to look locally.
4. Take a six-month course in machine learning.
5. Take a twelve-month course in data science.
6. Start tutoring local schoolchildren or JEE aspirants.

Now you can reassess your priorities and reduce the choices further. You can seek advice from a variety of people, not merely on what to do but on what each choice implies in terms of possibilities. You might then consider an amended list, such as

a) Take the job but keep looking online. (Recognize that looking does not guarantee finding.)

b) Go to a metro city and look there. (Others did it before you. You may find something. Or not.)

c) Take one of the two courses (machine learning or data science), and maybe supplement your income by tutoring local schoolchildren in the evening.

This smaller set is better. You can choose one from here. We will turn to that choice soon.

The above was for one set of circumstances. Now suppose your circumstances are different. Your family is financially strong, you are young and patient, and you believe you are actually very good at taking competitive exams. You just happened to study in a lesser-known college in a small town, and have not found a good job yet. And you are not presently interested in starting a business. Then you may eliminate a different set of choices from the list of thirteen. Maybe you will consider

1. Take a six-month course in machine learning.

2. Take a twelve-month course in data science.
3. Take GATE coaching until next year's exam date.
4. Start preparing for the UPSC.
5. Consider other exams, like the GRE, the GMAT, or the CAT.

Now, you can reassess the shorter list. The six-month course in machine learning presents a low entry barrier. Too many people can take it, so its benefits may be small or short lived. The three exams in item 5 all require somewhat similar skills, but they differ from those needed by GATE. Unless you have strong reasons to believe that you are special, item 4 has a low probability of success and may cost you heavily. These considerations may prompt you to reduce the list further to:

a) Twelve-month course in data science, with evening preparation for GRE, GMAT and the CAT.
b) GATE coaching for next year's exam. Full time.
c) Preparing for the UPSC. Full time.

Once again, you have a manageable smaller set to choose from.

You may have noted that both my examples of smaller sets to choose from have three items. Why three? Why not two or four? Because two is simpler, and four is harder than three.

With two options, i.e., binary choices, you can often boil it down to a choice based on principles. For example, do you value job security over scope for growth? Or, as an alternative principle, do you value autonomy over salary? With three options, such pairwise comparisons can sometimes be unresolvable. Remember the famous rock-paper-scissors game, where no option is better than the other two. Paper wraps rock, rock blunts scissors, scissors cut paper. With three options A, B and C, you may find that A seems better than B, B seems better than C, and C seems better than A.

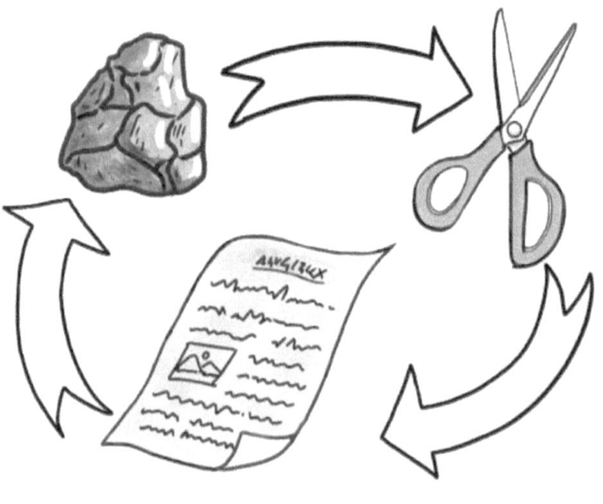

Choosing one out of three can be difficult.

Of course, in some situations you may be left with four or more options, and you still must choose as wisely as you can.

In this way, in many complicated situations, you might first discuss things with people, read a bit, go by what you have heard, and make a tentative list of possible options. That initial list may be long. Choosing the best option is then nearly impossible, so it is easier to identify some relatively poorer options and discard them. This is the first step: use heuristics to simplify things, to leave fewer choices on the table.

The next step is to reassess the options that remain, and maybe combine or restate them, to make a shorter list. At this stage, you can again get some external inputs.

Then, you must make a choice. If a superior option has become clear, you choose that and the matter is closed. If no option is clearly superior, you must introduce additional criteria. Your estimation of the probability of success of each option, if applicable. The range of future possibilities that each option leaves open (remember, life is high dimensional). Your emotional response to each option. The effect of each option on people you care about. Be *aware* of the criterion you are using, to avoid blaming yourself later.

Finally, commit to the choice you have made, move forward with faith, and – if things work out badly – try not to look back with too much regret. You took the best decision you could, given the information you had at the time.

That concludes discussion of my second suggestion, which is really a strategy to choose one option from among many different ones. The strategy involves a sequence of steps, summarized here: list your options, simplify using heuristics, shorten the list, be aware of the criteria you are using to make your final decision, commit to it, and – if things go poorly – do not look back with regret.

The third suggestion is to avoid inappropriate comparisons. Such comparisons are an inexhaustible source of grief.

It is clearly inappropriate to compare yourself with others whose journeys are very different from yours. It is instead more useful to see if your own journey can be improved.

Your journey, your battles, your successes.

I have a cousin who grew up in a small town. There is a railway station there, but it has only one platform. She stood first in her class in her school. She studied biosciences in a small college in a nearby city. Then she went for her MSc to a better university, one far from home. There, some of her classmates, who studied less than she did, performed better on competitive exams. They also knew useful things that she did not know: things that were not in her textbooks. I told her then that they came from different backgrounds than hers; and that she should focus on her own journey. Eventually, after a small job in Bengaluru and a short phase in Delhi, life took her to Switzerland. There she enrolled for a PhD. She was often amazed at how much some of her European fellow students knew. But she focused on her own journey. Eventually, she got her PhD and got a good job with a big pharmaceutical company. She kept moving forward, and the comparisons with others were eventually irrelevant. Her journey, her battles, and her successes – only these remained with her.

Another kind of inappropriate comparison is with success stories where we do not identify the appropriate *representative set*. Then we make basic mistakes in assessing probability, and take pointless risks. I discussed probability earlier, especially in the chapter on failure and persistence. However, the idea of a representative set is important enough to discuss separately and at some length.

Suppose, for example, that you are a village girl from an ordinary family. You read in the newspaper about some other village girl who did extremely well on some national competitive exam. You think that if she can do it, then you can do it. But that is emotional, not rational.

What is your representative set? In other words, which group of people can be said to be like you? Is it all village girls? Then your chances of matching that other girl seem negligibly small: one in millions, because most village girls will never match her. Is it all village students, boys and girls combined? Then your chances are still negligible.

You look closer. Maybe you find that the successful girl got good marks in high school and college, and enrolled in a well-known coaching institution. Perhaps you, too, are such a girl: good marks in school and college, and enrolled in a well-known coaching institution. This is a more selective representative set. Now it seems that your chances are better. Maybe one in some hundreds.

But maybe your representative set is even more selective. Maybe you find that this year was that successful girl's second attempt. Maybe you too have made one attempt so far, in which you were on a waiting list but did not get selected. Maybe you have continued to practice steadily since that first attempt, and have improved. Then maybe your chances are (I am guessing, of course) something closer to 1/10. Getting the exact number is impossible. But a rough estimate of where you lie can be made if you have a rational sense of your proper representative set. Without such a representative set in mind, you are flying blind.

Consider some other examples.

Suppose you are a student from a small town, but from a school which has excellent results year after year. Many former students from your school have done very well in their careers. Then the ex-students of that school can be your representative set. You, too, can expect to do well in your career.

Suppose you come from a business family. You have fifty family friends who have their own successful businesses. Then, if you start a business, your chances of success may be very good. But if you try for the UPSC, then your chances may be poorer.

Make a dry, unemotional assessment of your representative set for whatever you are going to attempt. This is not to say that you should not attempt it, or that you should not dream for big things. Statistically speaking, if you work hard, you can hope to do better than average among your representative set. But it seems statistically unlikely that you will do *much* better.

A businessman I once met told me an interesting story. He has a small factory that makes refractory bricks and exports to international markets. Many of his employees are poorly educated and have low salaries. He said that, in order to help his employees, he had offered them a career scheme for their children. He would sponsor the child for a certain level of study, and then ask the child to work in his factory for a certain number of years. Then he would sponsor them for the next level of study, and ask them to work for him again. And so on. It was a long, but steady, climb to good things. It turned out that very few of his employees' children seemed interested in his offer. Maybe they had unrealistic assessments of their own representative sets. Maybe they wanted to try things that offered faster growth, but with lower probability. Statistically speaking, I thought his offer was good. By this I mean that if 100 young people from such a set took up his offer and 100 others did not take up his offer, the first 100 would be better off on average. Several of his employees' children will probably end up worse off by not accepting his offer.

Of course, with each success along the way, your representative set can change[38]. Suppose you come from a small village, and enroll in a city college. Your representative set changes. Later, you enroll for another degree in a nationally renowned institute. Your representative set changes again. Maybe, from that institute, you get a job with a big company which offers excellent career prospects. Your representative set changes again. In that company, you build a reputation

---

[38] Probabilists may note that I am avoiding speaking of Bayesian priors. The distribution of attributes in the representative set is the prior. The sharper the representative set, the narrower the prior distribution.

for ability and character, and you move up into a smaller representative set: one that the upper management begins to watch. Maybe you then move to a higher position in another company. And so on.

A rule of thumb may help. Think of your career as progressing in stages. At each stage you try to do better than average within a reasonable representative set. That success gives you a new representative set. At the next stage, you try to do better than average in that new representative set, and so on. How much better than average is good enough? I suggest you should be satisfied if you do better than 60% of the people in your representative set. If you are more ambitious, try for better than 70%. For context, note that for a normally distributed variable, better than 70% of the population is slightly above the mean plus one standard deviation[39].

This last picture I have presented, of slowly but steadily raising your goals as you achieve higher levels of ability or success, matches something I tell my

---

[39] Years ago, when I was at IISc, a mechanical engineering graduate from a local college came by, looking for a project position. I asked him what his department's batch strength was in his college. He said it was 105. How many of those 105 got job offers while still in college? Four. No wonder he did not have a job. His classmates were his representative set. If his class rank was 31, he would have done better than 70% of them. If 30-35 people got jobs, he could work hard and hope to get one too. If only 4 of them got jobs, who knows what the stories behind those were? Maybe their parents had contacts, or they were extremely lucky, etc.

students: *Your aspiration must exceed your ability.* Not by too much, because that requires extreme effort and may lead to disappointment or burnout. And not by too little, because that is boring. Something realistic, but not trivial. Enough to keep you interested, engaged, and happy in your pursuit of happiness.

This concludes my third and final suggestion in the matter of looking forward in life: have a realistic representative set in mind, and try to do better than average compared to that set. Climb steadily, without disappointment or burnout.

Exercise 16.

Time for an exercise.

As engineers, we think of systems as existing in space and time. For this exercise, think of yourself as existing in space and time as well.

By space, I mean your world as it is today. In it, you are young, and you see your family, friends, colleagues, employer, etc., as they are today. Here and now is where you take life changing decisions.

But time will pass. If you are 25 years old today, there is a 35-year-old version of you waiting 10 years later in time, and a 45-year-old version waiting 10 years beyond that time, and so on. Some decisions that you take today will affect those future versions of you. You are their agent now, and they have neither voice nor choice except through you.

Many years ago, after I just graduated from college, I got a job in a factory. The work was good, the pay was nice, and I had a motorcycle that I rode joyfully fast in the way young people do. Every morning I went to the factory and punched a clock, and every evening I punched the clock and came back, feeling just as young as I did in the morning. But a feeling grew in me that one day soon I would come out of the factory as a 60-year-old man, not knowing where the years went. Some time later, I went abroad for higher studies, and my life was transformed. No regrets.

There is a trap, though. When you think of a time 40 years from now, you may imagine yourself wearing nice clothes, looking good, smiling amidst your family members, driving a beautiful car, and living in a beautiful home. That is not a plan, it is a mutual fund advertisement.

Your career is not how you look, but what you do and how you feel over the years when you do it. If a choice you make now is likely to make you unhappy in the future, at least be aware of the possibility.

CHAPTER 17

Later Challenges

In the early years you work hard, sharpen your skills, and take on greater responsibilities. That much is good. In the long run, however, other issues emerge. In this chapter I will write about time, money, burnout, and obsolescence.

First, let us talk of time.

As you make progress in your career, you may find that you have more and more work, e.g., with tighter deadlines, or because fewer people are allocated for the same amount of work. You may face more complexity in your work, e.g., working with many groups on varied projects with widely different time frames. You may find yourself having to meet deadlines on the work of others, in a managerial role. You may see that your team is overworked, and find yourself unable to help. You may find that family duties and health issues come into the same day which was earlier fully free for work.

Time is the ultimate limited resource. You cannot lengthen the day. You can only squeeze it harder. For many people, a career phase comes when the day is not long enough.

When you are young and have too much work, you may neglect your family and your health, and you may work late into the night. Sooner or later, however, your family's needs will be more urgent, your health may give you warnings, and the hours taken from the night will have to be repaid the next day. You will have to learn to manage your time.

By managing your time well you can extend what is possible, reduce damage to yourself, and be resilient for more years. There are many books on managing time. I offer two simple principles that engineers may appreciate, which I will also discuss within a design context in a later chapter. These are *sufficiency* and *efficiency*.

By sufficiency I mean here that you do enough but not much more. For example, if you are machining a component whose dimensions are supposed to be between 97 and 103, and you find at some stage that your component's dimension is 102, then *stop*. When a routine report is good enough for the purpose at hand, then submit it. You could spend another night polishing it, but it may not be worth it. And so on.

In the workplace, excellence is usually rewarded with some combination of more work, more praise, and more money (in that order). Watch your representative set, i.e., the people you will be compared with. If you are dramatically outperforming them, then either you are extremely talented (which is good), or you will wear yourself out (which is bad).

Obviously, I do not recommend either laziness or lies. You must do an honest day's work. But you should not work steadily on routine things at your *maximum* capacity any more than you should drive a car with the accelerator pedal on the floor all the time. Such all-consuming labor both wrecks your health and makes it impossible for you to think about things and extend the range of your skills.

After sufficiency, consider efficiency. Here, I mean developing your own ways of doing things quicker and better, in ways that leave you less tired.

Watch people at work, especially the ones who get more done and seem less tired. Factory workers making identical components on identically programmed machines can have different productivities at the end of the day. Two people writing similar ten-page reports on similar things may take different amounts of time. One person I knew, years ago, tried to look at each piece of paper that crossed his desk no more than once. Perhaps he developed memory tricks to help himself, but he surely did things fast and remembered more than others. Another person would start his day with a few minutes of quiet, make a list on a piece of paper, and carry that in his pocket. He stayed calm through the day, and got a lot done.

You may find that small exercises to improve your focus at key points, or working at certain times of the day, or taking a break every half hour, etc., may improve your efficiency. Look for ways of doing your work more cheaply in terms of time, effort, fatigue, or whatever it is that limits your workday.

Let me summarize. The world has an *infinite* capacity to absorb your work. You want to remain working for many years. Watch your representative set: do not try to beat all of them all the time. And whatever it is that you do, think of ways of doing it that leave you less tired.

Next, let us talk about money. Money is necessary. It is why people work. Having more money is usually considered better, which is why students starting their careers tend to prefer jobs with higher starting salaries.

There is a lot of advice available on money management. I just observe that two things about money can be unpleasant: (i) having much less than you used to have, and (ii) having less than you need at some critical time. The first can happen if you get laid off in mid-career[40], after becoming used to spending a big part of your former salary. The second can happen due to emergencies or poor long-term planning.

Being laid off is less likely if you have built a lasting value proposition. That is the main theme of this book, and I hope that you will indeed have a stable career for many years. The financial damage from emergencies can be reduced to some extent with insurance policies[41]. As a general principle I suggest that good financial planning is something that makes

---

[40] I know a few people in their fifties who studied mechanical engineering, went into IT, somehow slipped a little, and are now struggling to find employment they like.

[41] Some insurance policies are better than others. For example, many Indians do not understand the difference between term policies and endowment policies when it comes to life insurance. Some are shocked to hear that with a term policy they will not *receive* anything when the policy matures. They need to read and think about this issue. A reasonable strategy is to be skeptical about advice from the person who is selling you a policy.

you *less* dependent on your salary as the years pass. This can be through some combination of frugal living, systematic savings, and sound investments. The importance of good financial planning is felt most keenly by those who did not do it and got caught in a bad situation.

The third topic of this chapter is burnout. Burnout is a state of mind resulting from sustained stress at work. It makes you exhausted, unable to enjoy your job, and less effective at doing it.

I think that if you work too long every day for too many years without enough emotional rewards, you risk burnout. If you work at a job where, even as you get better at it, others around you start getting laid off, you risk burnout.

On the internet one can find several people from advanced engineering jobs in, e.g., electronics or software, who write that the stress level is too high, the working hours are too long, and the risk of being replaced by young people is high as well. Some of these people are in California, thought by many to be a place that offers unlimited opportunity. Perhaps the wage rate pyramid has caught up with tech jobs in California as well. As trained people kept flooding into California, at some point the job applicants were too numerous. Soon, workloads increased or, in other words, the wage rate decreased.

Yet, not everyone becomes burnt out. Perhaps you will not be. If the work you do is sufficient, and if you are efficient, and if you manage your money well enough that after a few years you do not *need* the job, then you are at lower risk of burnout.

Finally, let us consider obsolescence. Your skills are obsolete if there is no longer much demand for them. Obsolescence of some skills is normal in some modern subjects like software engineering, where one just keeps learning new things. But there is another way to become obsolete: a way that is harder to defeat. This is when the *worker* becomes old and slow and expensive; and younger workers claim the job. As a young worker, you may happily take a job from an older one (and why not?). Several years later, when you are older, the same may be done to you. I think the only way to escape this fate is to develop skills in other types of work, like moving from technology to management. But not all will escape. Thinking and planning one's path are better than drifting into obsolescence without being ready for it.

Age is not easy to defeat.

I was once speaking with a computer science graduate of IIT Kharagpur, some years my senior, now a successful entrepreneur in California. He said that what computer science or biology professors knew fifty years ago is now known to high school students. In contrast, many aspects of mechanical engineering have not changed at all. Mechanical engineering is therefore almost *static*, which he seemed to think was a bad thing. However, if the subject moves slowly, then obsolescence may come slowly as well. I have a cousin, a mechanical engineer from Roorkee, who has had a very good career with Bharat Petroleum

(BPCL). He once told me, "Young recruits cannot match what I have learned over 25 years. If I had been a computer science student, every year I would be struggling to learn new things that the new recruits already know."

The dilemma is clear, but the optimal solution depends on you. On the one hand, people who studied mechanical engineering may have earned only moderate wealth. In contrast, many people who studied computer science and did well are now rich, and may have moved into managerial positions. However, some techies are being laid off, and others are taking early retirement. In comparison, in the older subjects, obsolescence is slower.

One possibility is to earn fast and early, and retire when the pressure rises too much. But if your work is part of your self-definition, then early retirement may leave you a bit lost. Instead, you may like to switch to work with lower pay, but with content that you enjoy. After all, children may ask each other, "What are your hobbies?" Adults often ask, "What do you do?"

In this chapter I have pointed out four issues that are serious and must be faced. In the next chapter, I consider principles for facing these problems using the wage rate pyramid model.

Exercise 17.

An exercise. This has a long time frame. Think about things you enjoy doing well. For example, think back and see if there was a subject or a project that you enjoyed.

If you went somewhere for an internship, did you find the atmosphere energizing, or depressing, or stressful?

I once tried a Google search for

career passion

and Google said it had about 480 million results. Try such a search yourself.

It seems to me that, in the context of careers, the word "passion" has been overused into meaninglessness. I have not used this word anywhere else in this book.

But how about interest? Enjoyment? Satisfaction? Without positive feedback loops from such feelings, I think it is hard to maintain high levels of effort over long periods of time.

Some people think that we each have some magical predetermined interests. I think otherwise. There are many things that each of us could potentially be interested in. When we do something well, and we *know* we have done it well, then our interest in it can grow. As our interest grows, we tend to do even better at it.

So, you can try to be aware of what work felt good when you did it well. If you are not alert, you may not notice. If you do notice, then your interest may grow. If your interest grows, you may do more of that type of work. And this is the key point: you may find such work easier to sustain over a long time.

Being aware of one's feelings is not as easy as it may sound. Let me offer a simpler example. I once went to a conference at a university in Scotland, which is famous for its variety of whiskies. There was a whisky tasting session. A local expert on whisky was pouring out small amounts of one whisky after another into a circle of glasses. As people smelled or tasted the whiskies, he would wait a bit and tell them something to watch for, and *then* they would notice and smile. Apples, he said for one whisky. Caramel, he said for another. It was fascinating.

Unfortunately, as you try out different types of work, no expert will appear to tell you which ones felt good to you. But if you remember to think about how your work makes you feel, you will know.

CHAPTER 18

The Pyramid Holds the Key

The wage rate pyramid model, with which I started this book, is abstract but powerful. It represents the truth, as far as I know it, of private sector employment.

The pyramid is not static. It contains many stories of change.

The simplest and biggest story is of large numbers of people upgrading their skills in the hope of moving up the wage rate hierarchy. This is why many young people study engineering, many engineers study computer programming, and many computer programmers study AI and ML. This story has limited scope, however.

When too many people acquire a well-defined skill set, the wages associated with that skill set go *down*. It may seem that the corresponding employers or company owners benefit from this. To some extent, they do. But they are part of their own pyramid, too. Other companies move in and cut away the available profits. With India's demographics, I think that upgrading skills for the whole population, while highly desirable, will not lead by itself to better wage outcomes for all. Let me summarize this as follows.

Systematic incremental career improvements for the entire population are impossible by skill improvement alone. That skill will simply slide down to a lower position on the wage rate pyramid.

For example, if most Indian engineers are indeed unemployable, then training them all to be more employable does not mean that lakhs of new and good jobs will magically appear every year.

The above limitation of large-scale incremental growth leads to the second story on the pyramid. This is a story of high aspirations with low chances of success. In this story many will try, but only a small proportion will move up in big jumps. The numbers succeeding will be small, so they will not cause a downward slide in the pyramid. Yet, because the people at the lower level are many, there is a business opportunity in training them for the attempt. This is the story of many low quality "English medium" schools in India, where poor families send their children in expensive acts of hope. It is the story of coaching institutions for the JEE or the UPSC exams, or similar things. It seems that the bigger the jump and the smaller the chances, the greater the business opportunity. Because the demographics are challenging and hope is limitless, I suspect there are other opportunities of this nature, waiting to be tapped.

I have realized that it is not a bad thing to coach large numbers of poor people for a low probability attempt at something good. As an IIT Kanpur student from a village once explained to me, there are not many career paths that lead out from his village. Coaching institutions have given people like him a chance to achieve something big. As someone who has had better opportunities, I feel humbled by that student's statement. Yet, if a million people take coaching for exams where fifty thousand will be chosen, it does seem tremendously inefficient. How would we feel about a headache medicine that only worked 5% of the time?

The principle carries over to many levels on the pyramid. If there are a million students preparing for the JEE, there is scope for helping them. But if there are ten thousand working people trying to learn AI and ML in the evenings, there is scope for helping them too: their numbers are smaller, but their budgets are bigger, and the business opportunity is significant.

The final, and most uplifting, story within the wage rate pyramid is the introduction of new work. This is what employs more people, raises the wages of those that are there already, and creates new opportunities for those who are yet to enter. Starting about one generation ago, the Indian IT companies brought in a lot of work from outside India, and that has helped tremendously.

In your career, you may think about finding new work to introduce into the pyramid[42]. I think the opportunities are many. Let me explain why.

First, the amount of wealth in the world is not a constant. It keeps increasing. For example, in India, more people have electricity, more people have cars, and more people can afford to fly when they travel. If there is wealth, then there is somebody who wants to buy goods and services. Perhaps these goods and services are not on the market yet.

Talking of the world, there is a remarkable dynamic playing out these days. The native populations of many rich countries are declining. The Chinese population is

---

[42] If Indian engineers develop superior skills and attract work from overseas, then they introduce new work into our local pyramid.

shrinking fast after a generation of their one-child policy. Now they have fewer young people to produce the next generation, and those young people do not want many children of their own. I visited Brazil once, and was impressed by the depth of their technical universities, but the population there largely speaks Portuguese. Putting these thoughts together, I suggest there may soon be too much work in rich countries along with too few of their own young people. Their own young people will take the work that they enjoy, leaving the rest for others. Some of that work will be in technical subjects, which are dry and have a long learning curve. Going forward, we may see much engineering work coming to India from overseas. If you want to profit from that later, raise your game now.

Second, if you are trained as an engineer, that does not mean your career will be in engineering. You have transferable skills. I have a friend who was trained as an engineer but did various things in his life including growing tomatoes and raising dogs, and seems to have settled on running restaurants. He runs them with precisely tuned processes so that his food does not drift in either quality or taste. He has enough time to travel into forests from time to time to photograph wildlife. He says his restaurants run well because he uses basic engineering tools like measurement and standardization, along with good management based on understanding people and what motivates them[43].

Recall the discussion of failure and persistence. India is a very large country with many different people who have needs and wishes that we do not know of. If you talk to people, have an imagination, and persist, it seems likely that you will find something to do that has financial implications: something with which you can carve out a place for yourself in the wage rate pyramid, long before others find out and try to lower your earnings.

Once, encouraged by my young colleague Shikha who was doing some outreach work, I visited a village about 40 km from our campus. The villagers grow tomatoes, among other things. In the hottest months, their tomatoes dry quickly and cannot be sold. Sometimes a dealer arrives and offers low prices, and sometimes the dealer does not show up at all. Yet, tomato prices in those hot months are high. The village does not have reliable electricity and, in any case, it seems impractical to refrigerate so many tomatoes until they are sold.

I had an idea. About 6 meters into the ground, the temperature does not change much through the year.

---

[43] Money is a poorer motivator than you may think. Read *Drive* by Daniel Pink.

It is close to the annual average temperature, which is about 25-27 deg C. That is like air-conditioning compared to the summer daytime temperature of up to 45 deg C. I suggested that if they put the tomatoes in crates and lowered them into a dry well (just short of hitting water), then the tomatoes would be in a cool and humid place. They could be pulled up later.

An experiment was conducted. Tomatoes left outside shriveled in a day, and tomatoes lowered into the well were good for a few days. I must admit the villagers did not seem inclined to immediately get up and dig wells. Perhaps they were not convinced, or perhaps our language and clothes suggested to them that there might be government funded schemes for such things.

But the experiment was promising. The area of the well is obviously much smaller than the area of the farm, so the land exists. Crop spoilage is expensive, so the savings exist. The physics is correct[44]. A well can be dug for approximately the cost of buying a refrigerator, offers larger storage space, and does not require either electricity or replacement. How can there *not* be an opportunity here? For saving crops to begin with, and

---

[44] I asked Chiradeep Datta (from chapter 4). He says that the temperature in underground pipelines matches these simple predictions. Hot fluid pumped in at above 50 deg C cools to 25-27 deg C within about a kilometer.

maybe even rural home cooling some time later? Some design and standardization, i.e., engineering issues have to be worked out.

The wage pyramid is not merely a structure that limits your career growth while you do routine things. It also offers opportunities for introducing new work. India is a large country with many unsolved problems. This book may be written for people who seek work. But, in India, there is also work that seeks people.

Exercise 18.

A long term exercise. Nehru spoke of a scientific temper. Consider an *engineering* temper.

Look for badly done work. Notice electrical switchboards mounted crookedly, with switches so close to sockets that if you plug something in you cannot operate the switch; walls that are not straight; pipe joints that leak; washbasins where the taps are so far back that you must lean forward at an uncomfortable angle; welds that are highly uneven; furniture handles where too-short bolts go in only to about a depth of 2 or 3 threads; steel screws put into aluminum where they strip the threads; plastic chairs that buckle or crack; curtain rod clamps that come off the wall; patchwork that remains lumpy and unfinished; permanent splatters of paint on walls that landed when an upper floor was being finished; fan speed regulators with 5 settings where 4 are too slow and the fifth is full speed; and thousands of similar things as you go about living. Where a 2 percent increase in attentiveness can give a 50 percent increase in quality, there must be scope for earning a living.

CHAPTER 19

Some Years Later: A Network of Insights

In an earlier chapter I suggested that your mind is like a large structure of things that you know, along with connections between them. Let us now look at a part of that structure, which consists of simplifying insights within some broad technical area. The technical area depends on your training and experience. My area happens to be in applied mechanics along with some topics in applied mathematics. You will find and develop your own area.

In engineering, we can face two different types of situations.

The first situation involves a big system where many experienced people have been working for a long time. An example is an engine manufacturing plant of a large car company. They already have a good design, one which is sophisticated enough that newcomers cannot master it quickly. They have processes which they know well and which have been in place for several years. They have a large engineering staff. It is not easy for an outsider to go there and quickly improve things. If you join such a company as a fresh engineer, that may be a very good career move. But you will start at the bottom, learn on the job, and slowly become useful over several years. If you are older and want to be useful to such a company as a consultant or other external person, then you will need either prior useful experience or a significant time frame over which you can learn specialized things. This chapter does not address such a situation.

The second situation is one where the product or process is relatively new and perhaps not yet familiar; or perhaps the plant itself is new or has been modified recently; or the engineering staff is small and inexperienced; and an unexpected problem has cropped up. Here, the people in the plant may not immediately know what to do. Quick progress may be made even by an external person who has broad exposure. In such a situation, a person with a large network of simplifying insights can be useful. This chapter is about such insights.

The idea of simplifying insights is abstract, so I will explain through examples.

More than thirty years ago, I visited a small department in a big company. This was a service department that did design and fabrication jobs which were not part of the company's main processes. They had just finished making a small lift with a scissor mechanism, powered by a hydraulic cylinder, intended for lifting a weight of one ton. There were various constraints like where it must fit, how big the table must be, and so on, to which they had paid attention. But the first prototype failed to lift the required weight. When we were talking, the designer asked me about a force analysis at different heights of the load. He was middle aged, liked graphical methods, and did not use computers. His trigonometry was rusty.

After writing some tedious equations with sines and cosines, I thought I should make a quicker work-energy check. Cylinder force times stroke *must* be bigger than weight times height. And it was not. No wonder the lift did not work.

Let us not get distracted here by two things that are obvious but unimportant. (a) The designer should have checked this, and (b) anybody who knows work-energy methods will find this trivial. These are not why I am offering this example. The designer made a mistake – but I have made other mistakes elsewhere. He was a reasonable person doing an honest day's work. He had designed other things using the same catalogs of products (motors, hydraulic devices, bearings, etc.), and they had worked. He may have been preoccupied with size constraints, bearing choices, specifications of welds, assessment of friction, aspects of assembly, and so on.

My point here is also *not* that we must remember to do work-energy checks: that is too specific and limited. My point is that a simple work-energy check is one among *many* things that you might use, depending on the situation. Clues from the situation will ideally trigger the right association or connection in your mind. You need a network of such simple things to draw on. They must be in place before you encounter the problem.

My second example is from a visit I made to a well-known company that makes switchgear. In case you do not know, these switch structures are quite big (a few meters tall). They carry currents of, e.g., 30000 amperes, and they have to be opened within 32 milliseconds or under two cycles of AC power at 60 Hz (India uses 50 Hz, but some countries use 60 Hz). There is a spring whose preload is a few tons. This spring, when released through a complex arrangement of small parts with other springs and tripping devices, drives a mechanism that opens the switch. Some design and analysis of the switch mechanism was being done by this company using multibody dynamics software. However, I explained to them that since it has one degree of freedom, the system's motion can be mathematically described as a single spring-mass system. The spring stiffness is known. The effective mass is the sum of the effective masses of each separate moving part, and for some of those parts some approximation is needed because of changes in configuration. However, with the effective mass idea, an Excel file can be used to check the approximate effects of design changes in the system. Some years later, I met one of the young engineers from that company at a conference. He told me that the effective mass idea with a simple model had been useful in designing a new product which they had brought to market.

Once again, my point here is *not* specific to effective masses we can insert into spring-mass models. Rather, from Lagrangian mechanics, I knew that with one degree of freedom and a given potential energy, all I needed was a suitable kinetic energy term to obtain a useful simple model. Although Lagrangian mechanics and Newton-Euler mechanics are formally equivalent, sometimes one approach yields better insights than the other.

In another example, I was working with a former student who was employed by a motorcycle company. He used a software package called ADAMS Motorcycle, including a rather complicated tire slip model called the "magic tire formula", to do simulations of motorcycles going at various steady speeds on circular roads of different radii[45]. Looking at the curves of steer angles and torques for different speeds and radii for given tire properties, it *seemed* that they were similar, but stretched and shifted in some way. It was not obvious how to resolve the similarities and differences. Then we noticed, using dimensional analysis, that the number of nondimensional parameters was small; and that one of those parameters had a small magnitude. So, the behavior could be expanded in a formal Taylor series. Those insights helped us to collapse all the curves, for different radii and different speeds, into one single curve (for motorcycle and tire properties held fixed). Once again, a simplifying insight: but this time, it was dimensional analysis and a formal power series expansion.

---

[45] This was part of a larger study of motorcycle handling and stability.

In the same company, I heard another interesting story. They wanted to reduce the weight of an aluminum alloy casting, and were trying to change its shape while retaining some key dimensions, under some strength and stiffness constraints. Their young engineers tried optimization software for some months, but had not yet made useful progress[46]. Then one evening, one of their senior designers was standing at a food stall, looking with fresh eyes at a shallow *dosa tawa*. He noted that a small curvature in the *tawa* gave it much greater bending stiffness than a flat design would have[47]. The component they were trying to optimize did indeed have a flat wall (a legacy from an old design). With that insight, the design team rapidly achieved their cost reduction goal.

About ten years ago, my friend and colleague Vishwanath was testing a scaled model of a ship in IIT Kharagpur's remarkable wave tank. The ship was quite different in shape from container ships. For container ships, he already knew that a dangerous roll oscillation can arise when the wave encounter frequency is approximately twice that of the natural frequency of

---

[46] Optimization software can indeed be useful. But, guided by human insights, progress can be faster.

[47] Those interested in stress analysis may know that the shallow curvature induces membrane stresses upon bending, and stiffens the structure dramatically.

roll oscillations (this is an example of a two-to-one resonance). He was testing his model for different wave heights, at frequencies close to 2 times the roll frequency (say, 1.7 to 2 times that frequency). Now a ship in water is a complex thing. The ship, if simplified and modeled as rigid, has six degrees of freedom; and the water has infinitely many. But I had faith in a single degree of freedom model with what is called parametric forcing, and for such a model the two-to-one resonance is well known[48]. I requested Vishwanath to stick to exactly 2 times the roll frequency, and to increase the wave height in small increments (there was some risk of flooding the measuring equipment, so he was cautious). At a somewhat high wave height, and to his surprise, the two-to-one resonance was clearly seen. I think he was thinking of what he knows about ships in water, which is a great deal. But I was thinking of the simple two-to-one resonance. If you are surprised by the two-to-one resonance, think of a child pumping a swing. The child kicks twice per oscillation of the swing, and so the forcing frequency is exactly two times the oscillation frequency. The same insight works in many places.

Several years ago, I visited Neeraj Biyani's factory (he has written a chapter in this book). They had a

---

[48] You may like to look up the Mathieu equation.

troubling leakage problem. Their beverage was being filled into pouches in their plant. They had changed the cap design to something more beautiful, which changed some of the hardware in the plant. The leakage was a consequence of those changes. They were getting about 1 leak in 500 pouches, which is high enough to shut down the plant. We eventually discovered that the leaky pouches had tiny indentations in their plastic mouths. These indentations appeared during filling. I had the simplifying insight that the transition from reversible to permanent deformation occurs over only a small increment in load. So, if we could just reduce the severity of the capping load a *little*, the problem would go away. There was also a simple analogy we thought of in terms of the difference between screwdrivers, which require grooves in the screw, and butterfly nuts, which are tightened by fingers. With help from clever plant personnel, we soon managed to reduce the load a little. The problem went away and has not returned (leaks are now about 1 in 25000, I am told).

Some years ago, I visited a power plant with my friend Abhijit, a civil engineer who has done a lot of field work in health assessment of aging structures. While we were looking up at the chimney (which I found surprisingly tall), he told me about some work he had done elsewhere to assess the health of one such chimney. The work involves climbing to significant heights on the outside of the chimney, and using special nondestructive probes there. I expressed some wonder at the heights to which he had climbed, and he said that he had not in fact gone all the way to the top. Approximately one third of the way up is where you have to be most careful, he told me. This is because, higher up, the bending moments from lateral disturbances are small, and so tensile stresses are small. Closer to the base, the compressive loads from the chimney's weight are large and so net tensile stresses are small again. Since tensile loads are of greatest concern where concrete is concerned, about one third of the way up the chimney is where it might be weakest. A lovely insight, although I have not been able to use it yet!

Let me finally offer an example of a time when an insight came a bit late. Fortunately, no harm was done, because the right decision was taken as a default. Two senior people from a company dropped by my office. In their plant they had a boiler (for non-mechanical engineers: internally pressurized shell, externally heated, filled with water and steam). Most of the boiler shell was fine after many years of use, but one part had developed a number of cracks growing from the inside. They had many crack depth measurements from the shell wall and showed me their data. Two cracks seemed quite a bit deeper than the others. Clearly, they needed to replace that portion of the wall. Their question was, should they do it right then, or in the next financial year? The technical problem is difficult. I did some basic strength estimates. I disregarded the small number of cracks that were significantly deeper than the others. I told those gentlemen that I was not sure, but maybe they could wait and watch a bit if they wished. As it happens, they proceeded with repairs soon. I am glad they did, because of an insight that came to me later.

The insight is this. The boiler is heated from the outside using flue gas, and its shell wall is hotter outside than inside. Cracks starting on the inner wall, slowly growing through the wall by a creep like response, would experience both higher stress intensities and higher temperatures as they grew. This could accelerate the creep response, potentially causing a sort of runaway growth in a few cracks that lead the pack. This might explain those two cracks that were noticeably deeper than the rest. I am glad that I did not give them a definitive answer, and I am gladder they went ahead with the repairs! I think they trusted their instincts, did not have a definitive answer, and took the safer option. There are many unknowns in engineering, and a key idea is to avoid risks that are not necessary.

I hope the point of my examples is clear. The aim is *not* to develop a list of things to know, or a checklist to carry. It is to have a network of *ideas* from the subjects we study well. These are fundamental ideas. They are reliable. They can lead to strong insights into specific problems.

Some people are better at providing such insights than others. My friend and batchmate Atanu thinks very clearly about electronic circuits (among other things). I remember him, as a student, telling me briefly how to think in simple ways about transistors and about operational amplifiers. To this day, almost all I understand of electronics is based on those conversations[49]. His insight into such things was not superior merely for mechanical engineers like me. Before exams, his classmates from the electronics department routinely visited him for clarifications and insights. They would stand in a queue outside his room, late into the night. He helped them one by one.

If you can develop such simplifying insights on your own, that is surely a bonus. But even if you merely *recognize* such insights and think through them, work out examples using them, and slowly make them a part of your vocabulary of ideas, you will be ahead of the

---

[49] Due apologies to my teachers, who tried. Atanu just connected better with his fellow students.

game. Wherever extreme sophistication is not needed, and problems have not already been deeply studied; wherever things are still a bit new and some quick advances can be made – there, simplifying insights can accelerate your progress and help you contribute.

Academically speaking, my taste for such simplifying insights has been amplified by exposure to my PhD advisor Professor Andy Ruina, whom I must acknowledge here. In the area of mechanics, the simplifying insights he offered would frequently amaze me and my fellow PhD students. He would watch technical presentations and say things like, "That part is like an Euler column buckling," or "This part of the curve is a Mohr's circle in disguise," or "Here's a simple two degree of freedom system that will show that same resonance," or "Think of this big part on the left as infinitely massive, and then you will see why it does that." He made it look easy.

But I do not have to be him. What I learned and understood from him is mine now. And I have built up some insights of my own. Insights about bending, twisting, stretching, buckling, resonance, friction, sliding, rolling, contact stresses, fatigue, plasticity, stress concentrations, statistical scatter, stresses in two and three dimensions, natural frequencies, mode shapes, harmonics, impacts, restitution, energy, momentum, orthogonality, eigenvalues, singular matrices, boundary value problems, and many other such things. In technical work, this network of insights is now a part of my crystallized intelligence. Occasionally, a bright student asks me, "How did you know that?" And I say, "It is not difficult. It just takes 25 years."

If you are 20 years old now, you may think that 25 years is a very long time. It is, but it also is not. 25 years from now you may still have another 20 working years left. It is not like you can stop these 25 years from passing, either. The only choice you have is to be, or not to be, alert for interesting simplifying insights. If you are receptive and curious, you will find such insights from time to time. The world is big.

Exercise 19.

For this chapter's exercise, let us play a mathematical game.

You can use a circle on a sheet of paper, a ruler and pencil, and a random number generator. Alternatively, you can mimic the construction with a computer program if you like.

Around the circumference of this circle, mark 10 points (equally spaced if you like). Number these points 0 to 9. This is your model of a world where there are 10 important things to know.

Now generate two distinct random integers between 0 and 9. Let them be n and m. If $n = m$, do nothing and discard the step. Otherwise, use your ruler to draw a line connecting points n and m. This concludes one step in the construction.

Repeat the above step several times. As the game proceeds, count the step even if there is already a line joining points n and m.

Soon, several points will be connected. Here, if point 3 is not directly connected to point 7; but 3 is connected to 2, and 2 is connected to 5, and 5 is connected to 7, then we say that points 2, 3, 5 and 7 are all connected to each other.

How many steps will you need to take before all points, 0 to 9, are connected with each other? The mathematical minimum possible answer is 9. Requiring more than 40 steps is extremely rare. My computer simulation (do yours if you like) suggests that the median number of steps is 14, and the average is slightly under 15. That is less than I thought it would be.

Thus, if you pay attention and try to connect the things you know, you may need relatively few pairwise random connections before they are all connected.

But 10 is a small number. What about 100 things, numbered 0 to 99? My computer simulation suggests that in this case the median number of steps is 249, while the average is just under 260. Even 249 is not big, if you think of the implications, and remember that what you learn over 6 years can be slowly connected and consolidated over the next 6 years. There is time.

If you stay alert for connections across topics as you go through life, soon most of the important things you know will be interconnected in a network. Please see the title of this chapter again.

CHAPTER 20

Some Years Later: A Philosophy of Engineering Design

I once spent some time working with a small group of mechanical engineers in a medium sized startup that was developing autonomous vehicles in the 100-200 kg range. Many people there were good with computers, algorithms, data, control, and electronics. In contrast, the mechanical design team was small. They were hardworking, but not yet experienced in developing new types of vehicles from scratch. They had worked hard, and already built a creditable working prototype.

Then it was time to take a harder look at possible improvements. That is when they involved me.

Several issues were identified. For example, variable load sharing between two driven front wheels may have been causing higher wear on the single rear wheel due to unnecessary lateral loads. The control strategy built into the drive motor had a discontinuity at the notional operating point, which had mechanical consequences. Their wire-based two-wheel brake system struggled to maintain equal tensions in the two wires, leading to asymmetrical braking. They had two bearings to hold the steering column, and those seemed a bit too close: they might fail early under lateral frictional ground loads. There were some tire slip issues due to improper weight distribution and also during turns. Their finite element analyses of two vehicle frame designs had slightly inconsistent boundary conditions, leading to unclear outcomes. They had alignment and frame rigidity problems which could potentially be eased with alternative structural designs. It was a good opportunity for me to contribute.

Working with these people was a pleasure. Problems were not difficult to diagnose, solutions were not difficult to devise, rapid initial gains were possible, and rapid implementation of suggested changes was often possible. All this, of course, was aided by the fact that they had an open attitude and were quick to respond.

In contrast, if I had been working with an established car company, I may have had to work harder and longer on a narrower problem. Gains may have been smaller and less clear. My suggestions may have been difficult to implement simply due to legacy or scale issues. Overall, I would probably have had less satisfaction and possibly less impact as well.

This chapter is motivated by things I have understood from my interactions with small companies like that autonomous vehicle startup.

As indicated in the above paragraphs and also earlier, there are two situations that we often face. One involves a well-known and complex system, with a large staff of skilled people who have been working on it for quite some time. The other involves a system that is less well understood, perhaps new or recently modified, without a large and experienced staff, and facing an unfamiliar problem.

These two situations, and how we deal with them, seem to fit into a philosophy of engineering design that may help you think about your career.

The first and nonnegotiable goal of engineering design is to make something that performs some function. It is only later that we want it to function better, faster, more reliably, more cheaply, and so on. The first goal can be called *sufficiency*. The second goal comes later, and can be either higher *efficiency* or higher *performance* depending on what is attempted.

Sufficiency is supreme. If we design a helmet, sufficiency means it sustains prescribed impact loads. If we design a frame for a billboard, sufficiency means it does not fall over in the wind. If we design a pen, it should not break or bend when we write; ink should be smoothly released at the tip and not elsewhere.

For many big, expensive, and mature systems, the sufficiency goal was met long ago. For example, if you join an aircraft company and work on jet engines, and there you focus on heat transfer aspects in turbine blades, then you are working on either higher efficiency or higher performance. It will be some time before you can start making useful incremental design changes.

A mathematical view of the above three design goals, namely sufficiency, efficiency, and performance, can be offered in the form of *inequalities*. Think of it as something like a cartoon: a picture of the essential idea.

Let us say you are designing a frame to sustain a load F. The frame will fail at some load P. Sufficiency means the inequality

$$F < P.$$

Higher efficiency may mean using a weaker or lighter frame. That means P is lowered while holding F fixed, without violating $F < P$. Conversely, higher efficiency may also mean we realize, after some experience, that we can load the same frame more heavily. This means F is raised while holding P fixed, again without violating $F < P$. Finally, higher performance might involve holding something constant, like material use or weight or cost, while increasing P. This obviously allows us to increase F as well, again without violating $F < P$.

No matter what we do, we cannot violate $F < P$.

In this way, sufficiency in initial design is equivalent to one or more inequalities. Later, raising either efficiency or performance means tightening the inequalities from one direction or the other, or constraining the problem further, etc., but always while continuing to obey our inequalities.

Over time, after initial easy gains are realized, tightening the inequalities can become a hard and specialized job. Sometimes the margins available become really small[50]. However, for much of daily

---

[50] More than 20 years ago my friend Rajiv, a mechanical engineering PhD, was working for a big company that made computer hard drives. The push to improve the electromechanical components within the hard drive was relentless, because that is what increased the capacity of the drive. One flexible mechanical component was given to contractors to mass produce, where the transfer function of the component was specified

engineering design work for small companies, and for many engineering aspects of consumer product design, the inequalities are not very tight. For example, imagine doing three-point bending tests for a large number of pen designs. The bending rigidities and breaking loads will most likely be widely variable.

In engineering work, we are allowed to use theories, calculations, experiments, approximations, safety margins, and prior experience, in any combination that lets us make progress. We are opportunists, not purists. In the initial stages simple calculations, approximations, and generous safety margins can serve us well: we are in the sufficiency phase. On older problems, where much efficiency or performance improvement has already been achieved, safety margins become small. Then specialized theories, sophisticated experiments, complex and precise calculations, as well as prior experience within the domain, all become increasingly important.

Think about it for just another minute. An engine must last for a certain number of hours. A truck can be loaded up to a certain rated capacity. A pump fills your overhead water tank within so many minutes. Inequalities, all.

---

by the control design team. Such constrained design problems with tight tolerances are found in super-specialized areas.

Perhaps you have a product that is noisy. You want to make it quieter by 10 dB. If your present noise level is N_1 and after design modification it is to be N_2, then you are really trying to achieve

$$N_2 < N_1 - 10 \text{ dB} \cdot$$

An inequality, again.

Why is this insight important for *your* career? There are two reasons.

The first reason has to do with the jobs you might get in your career. The premise of this book is that many engineers may not get standard engineering jobs in established companies to work on improving mature systems. For the relatively few that do get such jobs, the traditional sophisticated discipline is already there. They will learn on the job and become useful over time. For the relatively many that will not get such jobs, the ones who will have to construct careers out of less traditional opportunities, I think reconnecting with this idea of sufficiency followed by efficiency and performance will help. Somewhere in your mind, you need to keep track of the inequalities that you must satisfy, and recognize the ones that are at risk of being violated. It will help you be more useful to more people.

The second reason has to do with the way we teach engineering. As we teach more advanced topics, and as our theories get more sophisticated, we bury ourselves in equations. And within our universities, I think that is how it *should* be. But when we go out into the world, we must reconnect with simple inequalities. Our equations are only tools to help us quickly achieve useful inequalities. If we fail to reconnect in this basic way, then the world has the right to call us impractical.

My friend and colleague Sonti, of IISc Bangalore, spent some years after his PhD working for a big company in the US. They had interesting tasks, such as reducing the noise level of old off-road machinery with big diesel engines. Initially, one of his non-PhD co-workers was skeptical about Sonti. He thought PhDs were impractical. Apparently there had been an earlier employee, another PhD, who was asked to develop some kind of model of the machine they were working on. That person, buried in his theoretical knowledge, struggled with the interaction between the machine and the floor. Academically, it is in fact a legitimate struggle. Is the floor rigid, or an infinite plate on an elastic foundation, or a half space? Is it elastic, or dissipative? Is it okay to think of the floor as a deformable hemisphere embedded in a larger rigid body? Can just a few springs and dampers be used to model the floor? The theory for each such case is fairly clear. The floor, of course, is none of these things. Engineering utility lies in picking the model that you can work with, in such a way that you do not get stuck or buried, so that you can draw conclusions which help you lower the noise level, i.e., achieve your desired inequality. Sonti's predecessor did not make such a transition from theory to utility. Apparently, he quit his job after telling his colleagues that he "couldn't model the floor." It was a joke within that group.

When we think of design (or even engineering troubleshooting) as a task involving inequalities, it has two advantages.

The first advantage is that in problems with many variables it lets us identify big effects, ignore small effects, and absorb the difference in a safety margin included within the inequality. If there are several inequalities, we can rank them in order of importance as well.

For example, when I was working with the autonomous vehicle team, they were interested in what happens when the vehicle tows a trailer using a horizontal link. Sufficiency requires a few simple checks. First, we estimate the pulling force, using the net weight and an estimated rolling resistance coefficient. Then, assuming no wheel slip or toppling, we find the required motor torque. Next, we estimate the wheel's normal reaction force using the center of mass location, use a reasonable but slightly low friction coefficient, and see if the available friction can prevent slip. Finally, because the rear wheel is driven, we check for toppling. For this, we apply the towing force, assume equilibrium while pivoting about the rear wheel, and compute the front wheel reaction to see if it comes out negative. Each inequality is easy to check in this case, and safety margins keep us out of trouble. A full simulation with ADAMS, for example, is overkill.

Next, we raise performance. What if there is a slope of a few degrees? We revisit our inequalities. The tilt changes the wheel reactions, which helps avoid rear wheel slip but aids toppling. The gravity component increases the towing force needed. This may be a good time to rank our concerns. If the vehicle topples backwards, it is embarrassing. We want a good margin on that inequality. If the motor torque is insufficient the vehicle may stop, but we may then lighten the trailer with not much damage done. If slip occurs on a slight turn, there is a chance of large sliding sideways, but our steer controller may compensate. How to account for some small forward acceleration? D'Alembert's principle says that it is the same as going up a small slope. What if the trailer is heavier? Similar considerations apply again. What if rear wheel slip is the main source of trouble each time? We could gain some advantage by using a downward-angled link which increases the normal wheel reactions, but eventually it will aid in toppling. What if we want to tow a still heavier trailer? Then we must consider ballast (e.g., optional heavy plates bolted on to the base).

Beyond a point, the answer does not lie in redesigning and recomputing. We must sell and deploy a few vehicles, troubleshoot them in the field, and learn more about them. We may learn good ways to use accelerometer data and offer warnings for overload, impending toppling, and even impending wheel slip. After some experience, our inequalities will be tighter and it will be difficult for a newcomer to immediately make improvements.

In this way, the first advantage of thinking of design in terms of inequalities is that it helps us focus on big effects and absorb small effects into safety margins,

The second advantage is that, unlike equations, inequalities have a direction. This direction can help us compensate for troubles that we cannot model, do not fully understand, and can only approximately quantify in terms of net effect.

In a motorcycle company I visited many years ago, I found a remarkable array of custom-made testing machines. They had a machine for lifting the upright motorcycle up by about one meter using a harness, and then dropping it. If I remember correctly, the frame and suspension were required to sustain 1000 such drops. Note that this method of testing replaces one poorly understood inequality (must survive on Indian roads) with an *ad hoc* but clearer inequality (must survive a thousand drops). Additionally, if the frame still fails when the bike is on the road, i.e., if customer complaints do come in, this method can respond by increasing the number of drops to 1500, or the drop height to 1.3 meters. In this way, a hard-to-compute inequality is replaced by an easily-testable inequality. In industry, such pass/fail testing is sometimes called qualification. But the principle of attaining sufficiency holds, and the qualification criterion is just one more inequality.

When I was fresh out of college, I joined Telco (now Tata Motors) in Jamshedpur. They had a large torture track for their trucks. The track was filled with staggered bumps designed to strongly load and twist the vehicle frame. Special drivers drove trucks, with standardized concrete loads, round and round that torture track. The empirical formula used was that a certain number of circuits (a few thousand) was equivalent to a certain number of kilometers (about one lakh) on Indian roads. If and when any significant structural change was made in the truck, it had to be qualified on the torture track.

I hope it is clear that these test-based inequalities, or sufficiency criteria, are not sacred. They are not derived from any fundamental theories of material behavior. They are, rather, empirical methods that can be used because we work with inequalities, and those inequalities have safety margins.

In the motorcycle company, they had several more machines for testing other aspects: the braking system, the stand, the shock absorbers, the chain cover, etc. What the test should be, how severe and extended the loading cycle, was itself something to be developed through some combination of theory, testing, and experience.

Sometimes, such indirect tests of sufficiency can be quickly devised to resolve isolated problems that require quick solutions. Once, when I was visiting a beverage packing plant, it was found that for one batch of packaging materials there was a slight change in the plastic sheet used, and packages were leaking. The next batch of material would be corrected, but some small process change was needed immediately to continue running the plant and to avoid discarding a large batch of packages. The portion that was leaking after filling was actually being sealed in the same plant, before filling, in a separate process. In that process, the component was clamped at a certain pressure for a certain time duration between heated jaws in a custom-built machine. A small amount of melting in the plastic occurred at the bonding interface, but the material remained mostly unmelted. We could easily change the jaw force, the jaw temperature, and the clamping duration. Other variables were difficult to change.

Since heat conduction was through the thickness, the physics was one-dimensional. A single nondimensional number was identified as important. This suggested that changing one parameter should be enough. Since there was not much flow in the original packaging design, the jaw force was not increased. Likewise, the temperature was not increased to avoid large scale melting. Consistent with a tiny change in thickness, it seemed most practical to increase the time duration of the clamping. But by how much? 5 percent? 20 percent seemed a bit high.

A test was designed. A person was chosen for supplying a fixed load (his weight). 20 filled packages, made at the usual setting, were tested first. The tester merely stood on them one by one, using one foot each time, placing his shoe in approximately the same way. All the packages leaked. Then the dwell time in the sealing machine was increased by 5 percent for 20 more packages, and the test was repeated. Only three of them leaked. Finally, the dwell time was increased by another 5 percent (i.e., 10 percent in all) for another

20 packages. None leaked. The 10 percent change was implemented.

The above experience may remind you of the Hindi word *jugaad*. I have mixed feelings about this word. At its best, it refers to using innovative methods with limited resources to achieve desired results, which is good. But sometimes it is used as a defense for sloppy thinking. I think, in this case, we were not sloppy. We knew which variables were probably important, we had a nondimensional number, we had an order of magnitude estimate of the change needed, we standardized our weight test, and we used statistics. Then we followed up to check that the problem was reliably solved. This was a mass-produced item, after all: reliability was important. The only seemingly unscientific thing we did was to use a man to supply his unmeasured weight as a load. But if you think about it, his weight over the half hour experiment was constant. If we had used a small block of steel to apply the load, it may have looked more technological but the improvement would not have been more valid. That the weight of the man was the appropriate load level, rather than the weight of a chair or a car, was based on our estimation of what a consumer might expect. A consumer who drives his car over the package will not complain if it leaks. If he stands on it and it survives, he should be satisfied.

Sometimes, *jugaad* is also used for work that is a one-off, e.g., a skilled street-side mechanic using small plates on the side to weld a rusted old bicycle frame[51] and help an owner extract some more use from it. While I appreciate such skill when I see it, much engineering work does not need such extreme resource-constrained ingenuity.

I will conclude this chapter on my philosophy of engineering design by reminding you that I have not written about *detail* at all. In fact, detail fills engineering. The number of components in a motorcycle or truck is large, and each has a process of manufacture. The assembly has a large number of steps. The autonomous vehicle group I interacted with had put in huge numbers of hours in designing the frame, with every component precisely specified, every joint and every bolt specified, with all the electronics and circuit boards and cabling and sensors and computing and algorithms working together, to get the vehicle on the ground and moving. And so on. As an engineer, you *will* deal with detail. That is merely not the focus of this chapter.

Stepping away from the detail and taking a higher-level view, engineers seem to think about systems in

---

[51] I knew the owner. I saw the welded result when she started riding it again. I was impressed. There was so much rust that I had thought it could not be done.

a way that resembles block diagrams. When we think of a large system, the blocks are bigger and each block contains more things under one heading. When we identify a problem within one block, we may break that single block into a whole new block diagram. There is room for experience and judgement there: what to pay attention to, what to retain in detail, what to club into a single blurry block because we do not think the main issue lies there.

Subsequently, however, individual tasks in design seem to fit my proposed view of sufficiency first, followed by improvements in efficiency and/or performance. And the mathematical cartoon I have offered, of thinking of it loosely as dealing with inequalities, has helped to clarify my own thinking. I hope it will help to clarify yours. Perhaps it may serve as a single key insight that helps you resolve a problem at work some day.

Exercise 20.

Let us try an indirect, thoughtful exercise.

People speak a bit casually about optimization. Optimization involves something that is somehow "best." But best by what criterion?

What is the best pair of shoes to buy? Do you value styling, comfort, level of arch support, quality of fit (e.g., some brands may fit your feet better than some others), water resistance, warmth in winter, longevity, or price? If you say all of them are important, how do you weigh them by appropriate amounts? When you go to a shoe store, do you look at some shoes and then say, "This pair is good enough and fits my budget?" *Sufficiency.*

How are students selected for admission by selective colleges like the IITs? Do they select the best and brightest students? The problem is that the best and brightest student (even if there was a reliable way to determine these qualities) may be hidden in some remote village with her ability not known to the world, or may not have applied for admission because she could not afford the application fees, or may have done poorly on the day of the entrance exam because of illness or a family tragedy. The college cannot look at all possible candidates and evaluate them with full reliability. So, it administers an entrance exam on a pre-announced date, runs many centers for the exam, and selects the top rankers. Far from ideal, but *sufficient* under its working constraints.

Suppose a new car model comes to market, and some time later customers complain that the front suspension has a flaw: it breaks when the car hits moderately sized potholes. The manufacturer realizes that the suspension does not meet *sufficiency*, and strengthens it. Is the suspension strengthened until it safely survives India's largest pothole with an infinitesimal margin? Of course not. It is just made stronger so that failures become rare. *Sufficiency* again.

What is the largest safe load which you can place on your laptop? The highest safe speed at which you can ride a motorcycle through your crowded neighborhood market? The closest safe distance to which you can approach a poisonous snake to look at its head, assuming you are not a practiced snake catcher? The largest stock market bet you can make without risking severe financial unhappiness? The biggest force with which you can write using your uncle's favorite fancy fountain pen? You may easily go through life without being able to answer these questions exactly. However, you might also propose some safe numbers. I suggest something like 5 kilos, 20 kmph, 8 feet, 2 lakh rupees, and 1 Newton. And then you may limit yourself to 2 kilos, 15 kmph, 12 feet, 20,000 rupees, and 0.3 Newtons. *Sufficiency* yet again.

If you enjoyed this discussion then you may like the design philosophy behind it too. Compress it still further, if you like, and write down somewhere: sufficiency before efficiency.

CHAPTER 21

The Fundamental Dilemma of Personal Ethics

In this book, I have asked you to be honest and reliable in addition to capable. I have recommended *ethical* behavior. Yet, it seems that the world has many successful people who are not too ethical. The contradiction must be addressed.

By *ethics* we mean a loosely defined system of values, of right and wrong, held by most people in a society at a given time[52]. Ethical rules are not clear and enforceable like laws. Unethical is not the same as illegal.

---

[52] Ethical norms evolve over time, but that is not important here.

Consider a hypothetical example. You work in a company. You, along with some young colleagues, are preparing for an exam to be held within the company. The person who does best will be promoted. Your colleague asks you which textbooks you are reading for the exam. If you tell him, you will lose some competitive advantage. You may choose to tell him, or you may not. The stakes are not big and the decision is not clear. If you do not tell him, it is not the same as stealing money from the company or physically harming someone.

There are some obvious plus and minus points, probabilistically speaking. If you tell him, he may get the promotion (which means you will not). But you may not have got the promotion anyway. Or, if you tell him, he may still not get the promotion and *you* may make a friend. Maybe he will tell you something useful later, on some different occasion. Then again, maybe he is just selfish, and will never help you back. There are many unknowns.

You may simply dodge his question, by laughing loudly and throwing your hands up as you head for the door. Or you may offer a minor lie: "I am too busy with my project, and am not getting time to study." It is not a legal matter, only an ethical one. As you go through life you will develop your own value system by which you will, or will not, answer such questions.

Maybe you will develop criteria for the level of prior friendship you require before you will answer honestly. More generally, you may be more ethical with some people than with others.

Clearly, ethical behavior is more complex and more personal than law. We are now ready to ask: why should an individual be ethical if there is no illegality, i.e., no fines, no jail time, no direct punishment from the system? That is the fundamental dilemma which this chapter aims to address.

The fundamental dilemma is an *individual* one. There is no dilemma for the collective. If most people are mostly ethical most of the time, then society obviously works better. When we call someone to repair our furniture, we expect them not to rob us. When we pay a shopkeeper in advance and ask them to deliver some things, we expect delivery. And each time we trust people in this way, we save time, money, effort, and annoyance. The collective is happier.

That does not explain why *individuals* choose to live ethical lives. If *you* decide to be ethical, what is in it for *you*? The matter used to be simpler than it is today.

In older societies, people were explicitly told what is good and what is bad, and some combination of social norms and religious beliefs prompted many people to remain ethical. If you presented the above situation (competitor-colleague seeking information) to a spiritual advisor from such a system, you might be advised to help your colleague. In such societies, *individuals* might conceivably have lived more ethical lives because their *spiritual advisors told them* to do so. If that is indeed what happened, it would address the fundamental dilemma I mentioned above.

But modern society is more fragmented. Many people leave their home towns, change residences often, change cities a few times, live more isolated lives, have friends scattered far and wide, and identify with online or virtual communities as much as they do with local ones. In some sense, religion plays a less powerful role in their lives; *and* when they do something unethical, people they care about may not be watching. Under such circumstances it is easier to do unethical things. Social media platforms also feed us stories and pictures of rich people living luxurious lives, people whose ethical standards are sometimes low. It is tempting to devalue the old advice on ethical living.

In other words, modern living has brought changes in our ethical compass. For example, in older times, greed was generally considered a vice. Today, greed is sometimes presented as a virtue. Sometimes greed is called ambition, and then it seems *obvious* that it is a virtue. This is the society we are in, and we are too small to change it. What we can do, however, is think about our value systems instead of letting them drift without our awareness. Let us be *selfish* – I accept that. Let us not be thoughtless, that is all.

In our evolutionary history, an individual who left the group tended to die. We therefore have an inherited tendency to identify with some group (an "us versus them" state of mind), and to even pay moderate costs for the privilege of belonging to our group. When we do something within the group that is not quite acceptable to the group, we see small gestures of negative feedback, like head shakes, disappearing smiles, and raised eyebrows. That negative feedback reverses our deviation and we return to the group.

Social media has changed things. Today, if we express a slightly deviant[53] opinion or interest on social media, the algorithms behind the platform actively feed us other items in the same direction of deviation. The deviation is not necessarily bad: it could be harmless. Maybe you like to look at videos of kittens playing with little balls: well, you will be fed more videos of such kittens. Maybe you are a woman who likes pictures of other women wearing colorful sarees. Quite harmless

---

[53] I say *deviant* in the sense of deviating from the mean, not necessarily either good or bad.

again, but your social media platform will respond. Maybe it is a little bit worse? Maybe you like to watch videos of people slipping, falling, and hurting their backsides. No problem, social media will find videos to show you. It could be even worse. Maybe you have shown some interest in videos of schoolteachers beating children. Your video feed will respond and see if your interest can be grown into an obsession. Each time we explore some slightly deviant behavior, social media platforms are ready and waiting with *positive* feedback. Positive feedback in such a situation can be destabilizing. If we are not watchful, such social media algorithms can take us further and further away from what our local group, or family group, or group of school friends, are doing. We can move steadily towards some virtual community with whom we feel a completely unreal, yet highly satisfying, kinship. If we proceed unmindfully along directions prompted by our social media feeds we will go, without alarm bells ringing, to places we did not choose. And whatever *that* group holds as commonly acceptable behavior will influence our personal ethical value system. In this sense, young people today may be experiencing an ethical crisis we have not yet understood properly.

Crisis is such a strong word, you might say. Why a crisis? Because, to the extent that ethics are social constructs, the society that is determining those ethics is now virtual, geographically disconnected, fickle, and rapidly evolving in response to the dynamics of our social media activity. And because, when we turn off our phones and walk in the streets or go to our offices, we must interact with real people in actual places, and those people may not have had the same ethical journey that we have had. The real people we meet every day, in terms of our ethics as well as their ethics, are strangers. The group that shapes our ethics is not the same as the group we live with. I do not think our evolutionary history has prepared us for this kind of society.

What should an individual do, in such times of unpredictable change? I propose that we must take a deductive approach, starting from basic principles.

What are the basic principles? I propose five of them. Number one: humans are social animals. We need to belong to a society, which is a group of people we identify with, whose rules we understand, who tolerate and even welcome us, with whom we share a common value system of what is right and what is wrong. Number two: that society cannot be virtual and ever-changing, it must be real and nearly constant over many years. Number three: developing such real connections may be harder now than before because of social media, and is certainly no longer an automatic consequence of living. Number four: as the unknown unknowns of the future come upon us, we may need a core group of people we can trust and who trust us. Number five: when we try to understand life, we use stories; and our stories are subject to survivor bias. When interpreting stories, we should not forget to take a rational statistical view of things.

Let us discuss implications.

Number one: humans are social animals. When you are young, you may think you do not need people if you can work from home, shop on Amazon, have food delivered to your door, have entertainment delivered to your screen, and use google maps to drive to various beautiful places. But recall our evolutionary history. Your genes have not changed because your data connection got faster. Hardened criminals in prisons are afraid of solitary confinement. If you think you do not need people, then you have not understood things yet. Maybe you have not had a chance to experience some of the pressures that life will surely bring. Life is longer and harder than you think.

Number two: you need a society that is real and nearly constant over many years. With that, we take our first step towards ethics. People have memories. When we interact with them on a repeated basis, their past behavior is known to us. If there are two shops side by side that sell the same things, and the first shopkeeper once cheated us, we may just develop the habit of going into the second shop. The first shopkeeper cannot force us to give him our repeat business, i.e., to become his loyal customers. The same happens in friendships and at work. If you want permanence in a group that takes you in as their own, you must behave honorably with that group.

Number three: developing connections is harder than before and is no longer automatic. How do you make old friends? By making new friends when you are young, and keeping them for many years. But, as discussed above, circumstances are transient, people move around, and they have virtual lives that may differ a lot from yours. The way to still make several good friends is to try with more people; do more good turns; be more generous, more strong, more reliable, more ready to show up and help. Much of your effort may not fetch a direct return, but some of it will. If you demonstrate high personal qualities to a hundred people, then maybe you will make five good friends. Here, as an older person, I offer a small secret: five good friends will get you through life. The problem is, you do not know in advance *which five* it will be. And so, you need to be decent to many more people than five. And there, you have taken your second step towards an ethical life.

Number four: as life's unknown unknowns come upon us, we need a core group of mutual trust. Here, I mean something beyond our closest friends, because the direction of life's attack is unknown. Maybe we will get laid off at age 45; or maybe one day our child will be home early from school, running a fever, when we are not there; or maybe our boss will get a dramatic promotion and choose two people to take upward with him; or maybe our spouse will fall ill and we will need someone with whom we can safely leave our children in the night; or maybe we will get in a bad car accident in a storm somewhere; or maybe we will just wish to help some young person by making an introduction to someone we know … Clearly, this extended circle cannot be as close as the circle of your closest friends. But this extended circle needs to exist, too. And how could you possibly develop such a circle unless you have a long track record of doing ethical things towards large numbers of people?

I think a strong, visible, and broad ethical approach towards a wide variety of people whom you meet and interact with is a shield against an uncertain future. Being ethical in the short term, many times, and towards many people, brings selfish benefits in the long term. By selfish benefits I do not mean a guaranteed prize, nor even a lottery ticket that some may win. I mean a sort of insurance policy that can only be bought in this way, over years.

You may have read the story of Androcles and the lion. It is an old story, and one with a simplistic moral. Androcles met the lion in a forest; the lion had a thorn in his paw; and Androcles took the thorn out. Later, both were captured, and Androcles was thrown to the lion. The lion recognized him and spared his life.

What do you take away from this story? To some people, this is a story of gratitude in the lion. To some, it is a story of good acts fetching rewards. To me, it is a story of probability and low-return investments. If you help one lion, and then later meet a lion in the stadium as its potential lunch, they will probably be different lions. The lion itself, in its majesty and lethality, is also just a storytelling tool: the same story would be less impressive if it was a rabbit who repaid help with a carrot. In essence, the lion is somebody who needed help, was given help, and later chose to be grateful. And just as Androcles's chances of meeting the same lion again were small, so also in life your chances of getting reliable returns from helping specific people may be small. But some returns may come your way if you help many people, and they may come in ways that count a great deal.

Finally, we come to my point number five: take a statistical view and avoid survivor bias.

Consider an example. There is a dangerous disease. If you get the disease, the probability that it will kill you on the first day is 25%. The probability that it will kill you on the second day is also 25%. On the third day, 25% again. But if you live past the first three days, then you will survive. So your probability of surviving the disease is, also, 25%. Now suppose somebody has got the disease, and has already survived the first two days. What is the probability of this person's dying on the third day? The answer is 50%. If this answer did not surprise you, then you are a better probabilist than I was at your age. If it did surprise you, then take it as a warning that you do not have reliable intuition for conditional probability. Survivor bias is ruled by conditional probability.

But first, to see that the answer is indeed 50%, let us avoid the cleverer Bayesian approach and take the simpler frequentist view. Imagine that 4 million people get the disease. One million people will die on the first day, another million will die on the second, a third million will die on the third day, and a final million will survive. When two days are over, two million are dead and two million remain. Of these two million, one million will die on the third day and one million will survive. There lies the 50%.

And so, oddly enough, at the start one's probability of dying on the third day is merely 25%, but if one lives through the first two days then the probability of dying on the third day rises to 50%. Of course, there is good news too. At the start one's probability of surviving is 25%, but if one lives through the first two days then the probability of surviving rises to 50%.

What does the above example have to do with long term ethical behavior? Imagine that you are systematically unethical, and your life is divided into 20 stages. The probability that the world will detect you, punish you, and discard you during any one of the first 19 stages is 5%. Now 19 times 5% is 95%, which is a high risk of failure. But if you survive these 19 stages, then you will win through to great material rewards. At the start of your career, your chances of winning unethically, by this example, are 5%. Now suppose you have survived through 18 stages. What is the probability that you will still be caught in the 19th stage? It is 50%. But, having survived through 18 stages, the probability that you will win is also 50%.

Sometimes people show me examples of people who almost made it, became rich and famous, and then got caught doing something wrong. Look up Elizabeth Holmes of Theranos some time. Sometimes, people offer me such stories as *proof* that dishonest people get caught in the end. I think they are wrong. Elizabeth Holmes almost succeeded. If she had played the game a little more cleverly, slowly leaked out stories of growing frustrations, and engineered a good public hand-wringing tearful admission of failure – maybe she could have kept a lot of money and stayed out of prison. But when we look at Elizabeth Holmes, I think we are looking at the last 2 stages of that 20-stage game. In those last two stages, in my example, it is fifty-fifty.

If you want to play her game, good luck to you. Do remember that she was an impressive-looking rich white American woman who went to Stanford. From her representative set, many people make lots of money.

If you are a typical young Indian engineer, people may not respond to you as they did to Elizabeth Holmes. Moreover, you are just starting the game. Your chances of winning by cheating, as per my example, are 5% and not 50%. Elizabeth Holmes *almost* made it. You may not go that far.

Exercise 21.

Let us close this chapter with an exercise in contemplation. Go over my five-point deductive list again. (i) You need a group of friends and well-wishers. (ii) That group needs to be somewhat stable. (iii) Building and sustaining such a group is not easy. (iv) The future is uncertain, so the group must be broad. (v) While planning, be alert for survivor bias.

Do you agree with my list? If so, I have succeeded.

As a separate matter, I confess here that being ethical is also rewarding. Ethical people take *pride* in being ethical. They *enjoy* it. But I have kept that out of the chapter above.

CHAPTER 22

Self, Workplace, World

Let me begin this final chapter by recalling some basic advice on savings and investing.

Start investing early. Think of your investing life as proceeding in three stages. When you are young, your savings rate is more important than your rate of return. Put away more money every month if you can, and do not worry too much about the return on your investments. In the middle years of your career, your rate of return is most important. This is when your investments grow rapidly, and you still have some years left to recover from setbacks. In the late years of your career, prefer safety and avoid risky investments. If your investments have done well, start spending some money on things that give you joy.

Why discuss investing in a book on building and sustaining a career? Because there is a similar three stage plan there, too.

Start learning early. Think of your career as proceeding in three stages. When you are young, the rate at which you learn is more important than your salary. In this stage you are building yourself. In the middle years of your career, focus on growth in the workplace. This can mean seeking positions of higher responsibility, more challenging projects, successful businesses you may have started, or perhaps even higher salaries[54]. This is when your career graph rises rapidly, and you still have some years left to recover from setbacks. In the late years of your career, aim for stable performance in your areas of strength. If your position allows it, spend energy on work that contributes to others.

These three stages do not have sharp boundaries. They are simply ideas to help you think about your career.

Many people do not, of course, follow the above three-stage plan. For example, *Quora* has many

---

[54] I say "perhaps" because my own salary is fixed, and my growth has been in the form of projects I take on, research papers I write, the careers of my students, etc.

people asking, essentially, "If I study subject X in institute Y, how much will I be paid?" They are not focusing on developing themselves, only on getting a rubber stamp that leads to a paycheck. I hope that in this book I have convinced you their approach is shortsighted.

In your early years, you must build yourself. Choose a fundamental discipline and master it. Find meaningful problems and learn how to solve them. Observe smart experienced people and learn from them. Take responsibility and be reliable. Do not worry if your salary is low, provided you are learning things that let your skills rise above the next generation of people in your field.

Then you enter your career growth stage. Growth, like life, is high dimensional. You can keep growing in your technical skills, which is good for designers and researchers. But you can also grow in experience, in your courage to take on large challenging projects, your ability to handle complex situations with poise and wisdom, and various other things. You can grow in your reputation for being a reliable person, so that senior people seek *you* out when they want a job done correctly in one shot. As I explained earlier, you may feel overworked or even exploited in the short run, but each time you do something difficult, your career grows too. If and when you leave one workplace to join another, your ability and confidence go with you.

As the years pass, career growth means bigger things. You lead large groups; or you take broader or bolder steps in your business; or you rise to a position of high authority in a large company, which requires you to cross or master multiple technical domains.

With conscious planning and some luck, late in your career you will be in a stable role where your crystallized intelligence is valued. At that time, you must seek greater meaning over money. After all, what is the alternative? Who wants to be an old man chasing his last flying currency note into the crematorium?

Meaning often comes from contribution.

You may find yourself mentoring younger colleagues, or helping to secure fairer treatment or better opportunities for lower ranked employees who need help. That is a valuable human contribution.

You may find yourself volunteering time to develop some new area of expertise within your organization, something like a thrust area of the future or a dramatically new lab facility. This will provide a space for the next generation to grow. That is a valuable technical contribution.

If you work for a large company that has social outreach programs, you may find yourself taking up work in those directions. Or, if you are a professor, you may find yourself writing a book like this.

So, in closing, I leave you with a three-word formula for the entire trajectory of your career: self, workplace, world.

www.ingramcontent.com/pod-product-compliance
Lightning Source LLC
Chambersburg PA
CBHW032007150726
47990CB00005B/1877